Flying with GPS

By Stephen N Clark

PILOT
Publishing

Published by Pilot Publishing

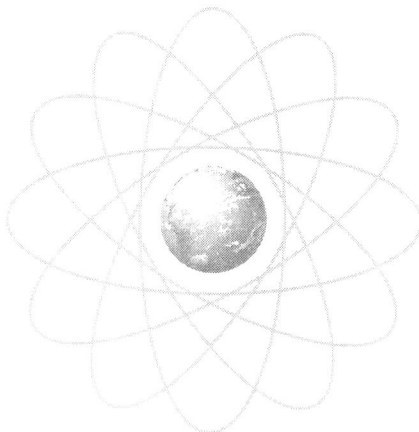

First edition November 2002
Reprinted with revisions February 2003
Text and photography © 2002 Stephen N Clark
Design and Layout by Dean Such

Disclaimer

Whilst all reasonable efforts have been taken to ensure the accuracy of the material and advice
contained in this book, the author, publisher and distributor take no responsibility for any action
or inaction resulting from its use.

Published by Pilot Publishing

Distributed by Pilot Warehouse

www.pilotwarehouse.co.uk

e-mail: pilotwarehouse@btinternet.com

Telephone: +44 (0)1442 851087

Fax: +44 (0)1442 851541

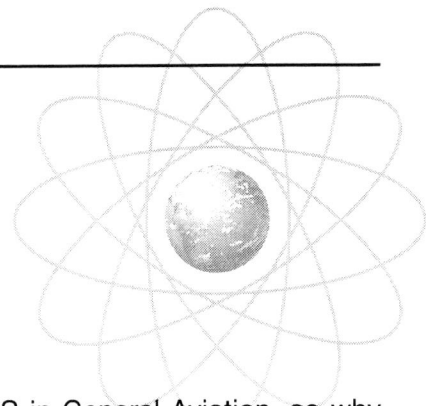

Why this Book?

A large number of publications are already available covering GPS in General Aviation, so why this book?

Pilots today face the problem that the books already available talk, either in very general terms about GPS as a tool with no real hands-on information, or take the form of reference books of co-ordinates of navigation aids and towns which must be tediously entered prior to flight. Alternatively, the pilot must turn to the GPS's manual that many find incomprehensible, explaining the features in a way devoid of practical context.

Another issue that faces new pilots is that, other than an almost passing mention under the JAR syllabus, use of GPS does not figure in formal training, and until quite recently they tended to be branded "the devil's tool" by many instructors. Yet the first thing many new PPL holders do is to buy a GPS to celebrate gaining their licence. Increasingly GPS panel fits are becoming standard in new aircraft often at the expense of radio aids such as DME and ADF, and units such as the Garmin GNS 430 are being seen replacing aging COM/NAV sets to comply with FM immunity requirements.

This book bridges the gap in information and training materials by introducing the use of GPS in a realistic context, whilst giving respect to the training in traditional methods received in study towards the PPL and UK IMC rating in order that pilots new to GPS can use the tool safely and be aware of its benefits and pitfalls.

Why Garmin?

Garmin is probably the most popular make of handheld aviation GPS in the UK (or "portable" GPS as Garmin call them), with all the units coming with easily upgradeable Jeppesen databases. Garmin panel mount receivers are becoming increasingly popular, especially the GNS 430 which includes COM and NAV radio functionality, often as replacements for older non-FM immune equipment. This book is also useful if you are trying to make up your mind which Garmin portable GPS to buy or have upgraded your model and want to quickly understand the differences and new features.

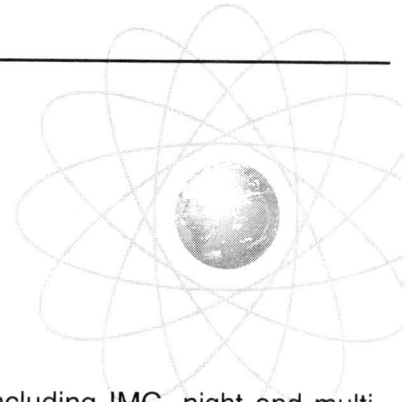

About the Author

Steve Clark has a PPL with several hundred hours experience including IMC, night and multi-engine ratings, and has flown all around the UK and Europe over the last few years both with and without the help of GPS. During his day-to-day work in the Computing, Internet and Media sector, he has always worked with leading edge consumer technologies and looked at the ways to get the best from them.

Following requests, Steve has lectured informally to flying groups and tutored individuals about getting the best from GPS and helping users understand everything from their basic functions through to complex route planning and even flying approaches.

Many students suggested that a book was in order, hence another reason "why this book?"

Acknowledgements

Many people have contributed to the writing of this book. Particular thanks are due to Tom Weiss, Philippa Dickinson and Linda Bradford who have "tested" the book out during its development and loaned some of the GPS receivers, and to Rod Brown, the CFI of Denham School of Flying for his encouragement and constructive critique. Thanks are also due to the owners of Piper Archer G-BOPA and Cirrus SR20 N147CD, that were used in some of the photography.

I'd especially like to thank my wife Charlotte who is training for her PPL and has helped immensely with testing out the book and supporting me while I wrote it.

Support Web Site

Technology can change very rapidly. This book is supported by a web site that carries additional information, updates, current web links to references in this book and errata. The web site is at **www.gpsbook.co.uk**.

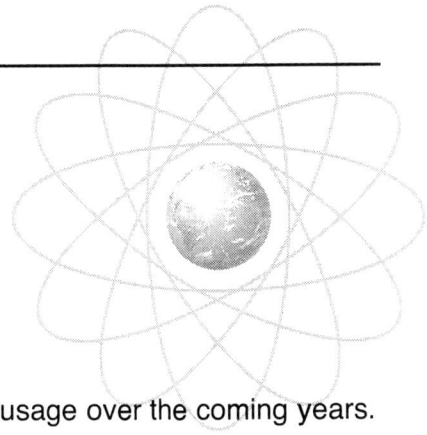

GPS Philosophy and this Book

Like it or loathe it, GPS is here to stay and is set to see increased usage over the coming years.

A lot has been written on the pros and cons of GPS and various pundits tend to take up a position at either pole according to the point of view they want to put across.

There is no doubt that:

- **A GPS will lose its signal, albeit occasionally**
- **A GPS will run out of batteries**
- **A GPS will fail for some other reason**
- **A GPS will distract inexperienced users from more important tasks (like flying the plane and looking out of the cockpit)**

There is also no doubt that when correctly used:

- **A GPS will help prevent you getting lost**
- **A GPS will improve your situational awareness**
- **A GPS will significantly ease flight planning**
- **A GPS will significantly ease cockpit workload**

It's strange that such strong debate seems to be reserved for GPS - one could write equivalent lists of pros and cons about almost any other piece of cockpit equipment - the difference may be that pilots are trained about what to do when they fail - you probably haven't been trained at all about how to use GPS effectively.

This book takes a balanced view and will enable you to use this valuable tool safely and effectively, avoiding its pitfalls.

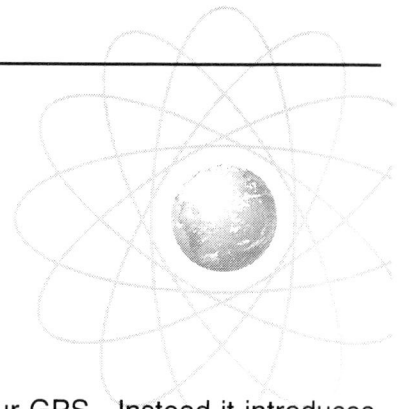

How the Book is Laid Out

This book isn't intended to replace the manual that comes with your GPS. Instead it introduces you to the GPS functionality through practical information and ground school exercises. This book contains most of what you need to know. Even so, you may wish to keep the GPS manual handy and use it if necessary to learn more about the features discussed.

The book is in several sections:

SECTION 1
- describes the Garmin portable range of GPS receivers and explains how GPS in general works.

SECTION 2
- looks at the basic operation of your GPS, including powering it up and finding your way around its "pages" of information and carrying out simple functions. There is some intensive "ground school" to get you ready for flight.

SECTION 3
- introduces GPS into the flight planning process. This builds on the VFR navigation training typically taught in the PPL syllabus, adding the benefit of GPS, without losing the benefit of traditional VFR techniques.

SECTION 4
- introduces features for more experienced pilots, including using GPS to improve situational awareness when using radio navigation techniques.

SECTION 5
- looks at how to maintain and get the best from your GPS including how to update its Jeppesen database and use it in conjunction with automated flight planning software.

SECTION 6
- Introduces the GPSMAP 196, which at the time of writing is Garmin's newest aviation GPS, and highlights the key differences with the other units.

SECTION 7
- builds on the knowledge learned earlier in the book, but focuses on the panel-mounted GNS 430, which is increasingly being found in many hire aircraft.

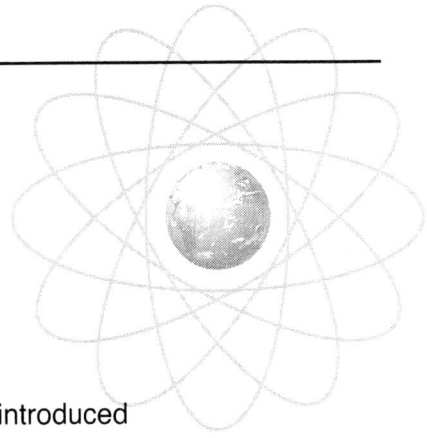

Information Panels

Periodically through the book alert panels are introduced
to highlight a significant point as follows:

Fact: Useful fact worth remembering or a key
point of note.

Airmanship: As with any aircraft manoeuvre, take
appropriate care prior and during the operation
under discussion.

Alert: Alerting you to an important fact.

Checklist: Consider adding this point to your
checklist or memorised checks for the juncture
of flight under discussion

Tip: Useful advice based on the Author's
experience.

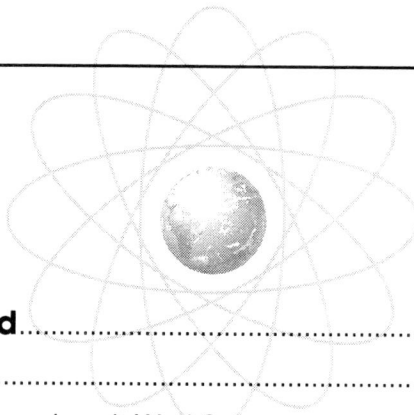

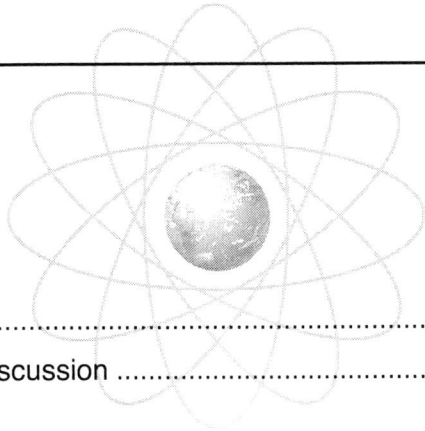

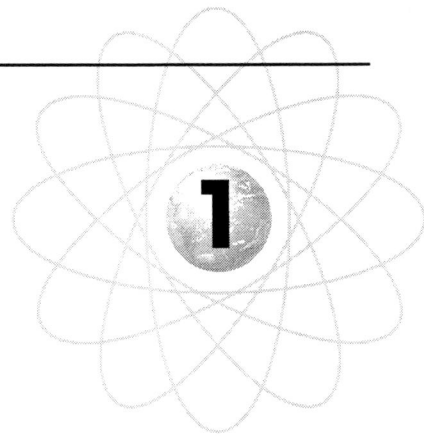

Getting Started

This section introduces GPS.
The first chapter looks at the Garmin yoke mounted range of portable handheld GPS receivers and provides an overview of their features. The second chapter examines the background of GPS and how it works.

Introduction

This Chapter introduces you to the Garmin range of portable GPS receivers.

This book discusses aviation receivers (as opposed to ones designed for land navigation). This is for two reasons. Firstly they are available with a yoke mount so that they can be positioned for safe operation within an aircraft, and secondly they come with a database of aviation waypoints.

> ⚠ Alert: Don't try to use a non-aviation GPS for serious aviation navigation. It is likely to be a distraction not a tool. Use proper VFR navigation techniques (and radio aids if qualified) instead.

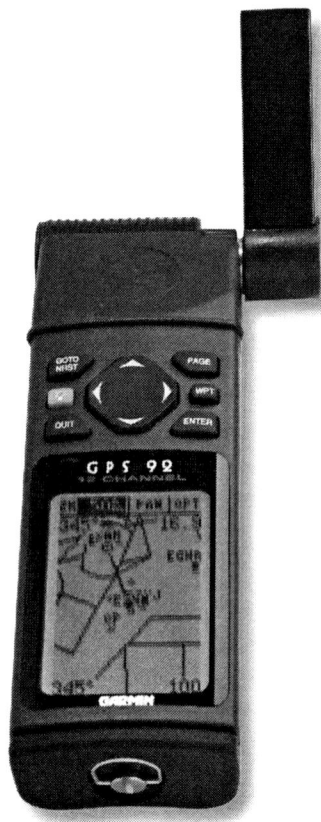

The GPS 92

The GPS 92 is now the basic entry point aviation GPS from Garmin and supersedes the older GPS 89 and 90 models.

It incorporates a monochrome LCD screen and 12-channel satellite receiver with a Jeppesen database, and can store 20 flight plans of up to 30 waypoints each.

The GPS 92 comes with most of the normal accessories including a handheld antenna, a remote antenna mount, batteries, a power/data cable, case, manual and reference card, though notably does not include a yoke mount in the standard package.

The screen is basic to say the least and the lack of a yoke mount as standard means that unless budget is at an absolute premium the slightly more expensive GPS III Pilot is recommended.

The GPS III Pilot

The GPS III Pilot provides an excellent entry point to aviation GPS, incorporating many features found in the more expensive GPS receivers in the range and is superior in many ways to the GPS 92.

It incorporates Garmin's 12 channel parallel satellite receiver and Jeppesen database. The screen is physically the same size as the GPS 92's, but has significantly higher resolution and rather than being monochrome it supports shades of grey so is more readable. It also has a base map underlying the aviation data.

The GPS III can be used either horizontally or vertically making it suitable for yoke mounting or "dash" mounting on the glare shield. Unfortunately only the "dash" mount is included as standard along with some velcro sticky pads. Two yoke mount options are available, the more expensive of which includes power and data cables.

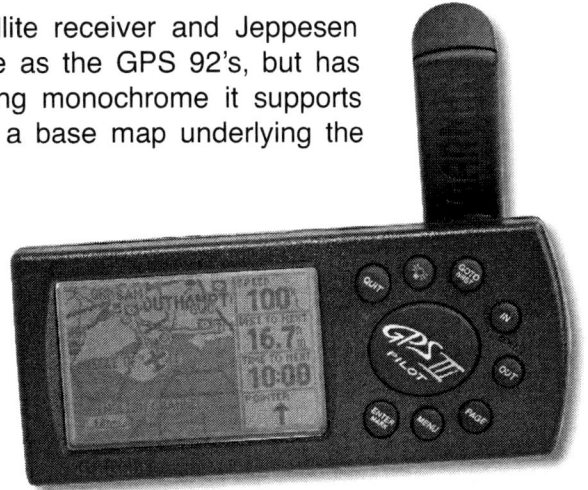

The GPSMAP 195

The GPSMAP 195 was Garmin's first truly powerful handheld aviation GPS and is dominated by its large configurable grey-scale screen which can be backlit.

It incorporates Garmin's 12 channel parallel satellite receiver and Jeppesen database, and also includes a base map of roads, railways and towns etc, plus final approach waypoints to assist in non-precision approaches. The screen is very readable and configurable (though unlike the GPS III Pilot it can only be used "vertically"). There are also a number of "bonus" features including an automatic weight and balance schedule.

The GPSMAP 195 comes with plenty of accessories including a remote antenna, yoke mount, battery pack, cigarette lighter adaptor, and carrying case as well as the manual and quick reference card. A rechargeable battery is available as an optional accessory though is relatively expensive and can be difficult to find. Unfortunately the GPSMAP 195 uses an uncommon antenna connection and unusual power / data coupling so other Garmin accessories can't be used with it.

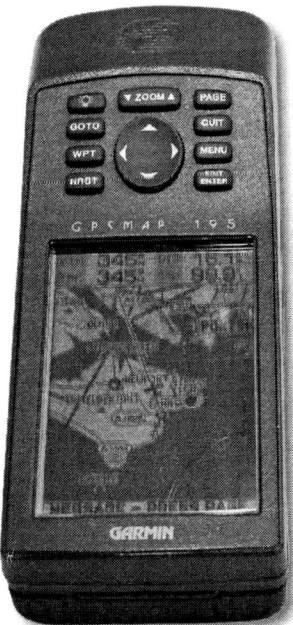

The GPSMAP 196

The GPSMAP 196 was introduced in August 2002 and is seen very much as a replacement to the 195.

This GPS combines aviation, land and marine based features in a single unit. Like most other modern Garmin units it incorporates the Garmin 12 channel receiver which tracks up to 12 satellites in parallel. It has fast acquisition times and supports Differential GPS (DGPS) and WAAS (see Appendix A).

The screen is a large grey-scale, though unlike the earlier GPSMAP 195 it has more of a horizontal aspect ratio than a vertical one. The map scrolling speed when manually panned has been significantly speeded up.

It shares a base map similar to that in the GPSMAP 295, which includes roads and can have a MapSource cartridge added so that additional detail can be loaded from CD-ROM. In land mode it can do proper vehicle routing (unlike the 295), though this is limited to the inbuilt mapping and doesn't extend to the more detailed MapSource data.

In aviation mode it has a standard Jeppesen database and offers the comprehensive facilities of most of the other Garmin aviation portables and includes some bonuses including the ability to enter specific aircraft details, a weight & balance and logbook facility.

Garmin have also significantly increased the number of routes and waypoints that can be stored and track logs can now be saved.

The unit comes with a good selection of accessories including a power cable, PC interface cable, yoke mount, glare shield mount and remote antenna. Unfortunately the carry-case is an optional extra.

Because the GPSMAP 196 was introduced just as this edition of this book was being completed a special section has been written to cover it - Section 6.

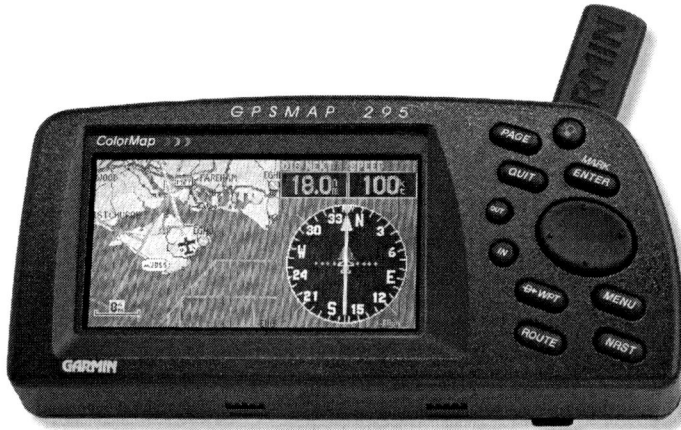

The GPSMAP 295

The GPSMAP 295 is at the top of the Garmin handheld range. It has a crystal clear display which is easy to read in almost any conditions, and which can be configured to suit individual tastes.

The receiver incorporates Garmin's 12 channel parallel satellite receiver and Jeppesen database, and also includes a base map of roads, railways and towns etc, plus final approach waypoints to assist in non-precision approaches. All of this is in colour making it very easy to read.

The receiver accepts add-on memory cards and is compatible with Garmin's MapSource range of CD-ROMs. This means that you can significantly improve the zoom capability of the base map, or add useful information such as restaurants, hotels and so on.

The unit also incorporates a non-aviation mode so with the addition of a MapSource CD it can even be used to assist in road-based navigation (though unfortunately not with the facilities that dedicated in-car navigation systems provide).

The GPSMAP 295 comes with almost every accessory you will need including "dash" mount for mounting on the glare shield (or in your car), a yoke mount, handheld antenna, remote antenna, cigarette lighter adapter, case, manual and quick reference guide.

This was one of the first handheld products to be WAAS enabled enabling a higher degree of accuracy (see Appendix A).

Introduction

This Chapter explains the basics of GPS, what it is and how it works.

GPS stands for Global Positioning System. The system consists of 24 satellites in orbit around the earth at a distance of about 20,000 km (12,000 miles), every 12 hours. 24 satellites are needed to ensure that at least 4 satellites are visible for an accurate position fix at any time anywhere in the World.

The system was created for the US Department of Defense starting as far back as the 1970's and finally being completed in the mid 1990's with the addition of the 24th satellite. Several more satellites have been launched subsequently which can be used as "spares" and provide additional facilities.

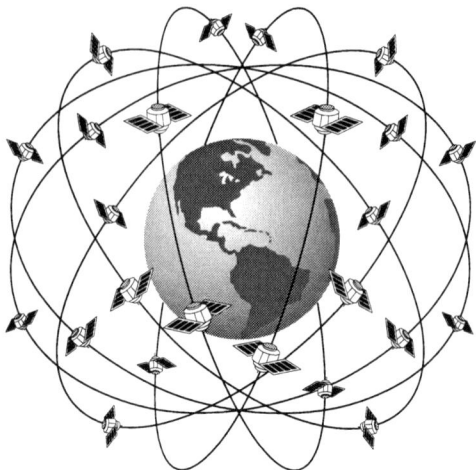

The constellation of GPS satellites

How Does it Work?

GPS works using a technique often called "triangulation", but more correctly called "trilateration", to determine the position of the receiver. Most PPL holders who have studied navigation will be familiar with the concepts:

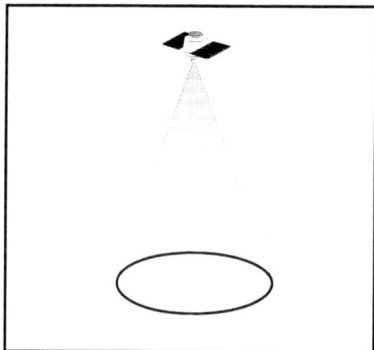

If you know how far you are from one known point then you can be anywhere on a circle...

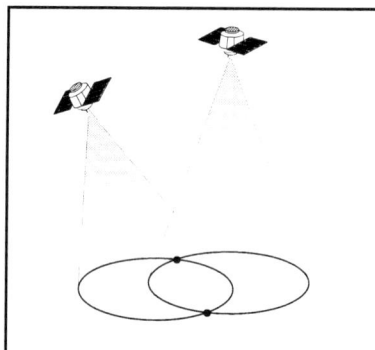

If you know how far away you are from two known points then you can be at one of two places where the circles intersect...

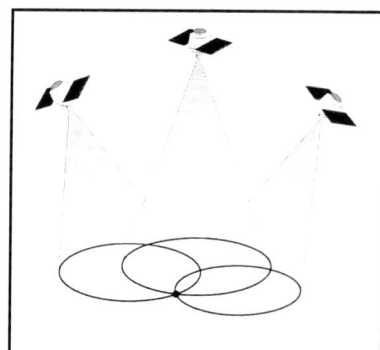

If you know how far away you are from three known points then you can be at only one place - where all three circles intersect.

What is GPS?

How Does GPS Work?

In the case of GPS, the known points are the satellites. They follow very accurate orbital paths and the receiver knows where they are supposed to be using an almanac. The receiver can get updates and corrections to its almanac from the satellite signals.

Of course you have to know the distance from the known points. This is done using timing. It is fairly well known that each GPS satellite has an atomic clock. Each satellite broadcasts an encoded stream called a pseudo random code that enables the receiver to derive the time the signal was sent. The receiver measures the arrival time of the signal from each of the satellites and is able to compute its distance from each one, since the signal will have travelled at the speed of light to reach the receiver.

To measure the time difference accurately would require the receiver to also have an atomic clock, which of course it doesn't. This therefore adds an additional unknown into the equation and makes the position slightly uncertain.

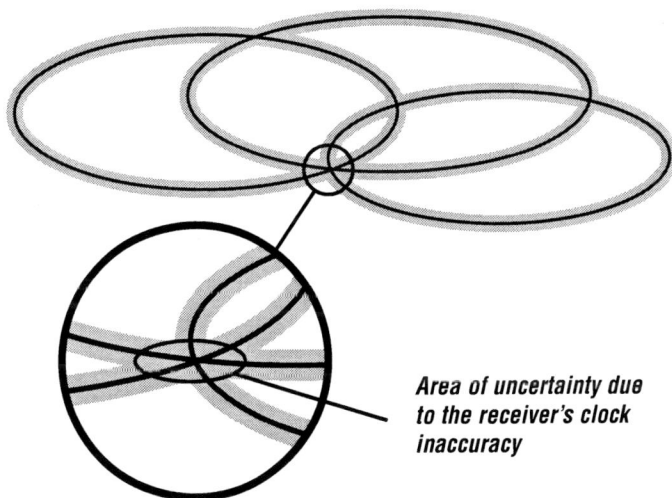

Area of uncertainty due to the receiver's clock inaccuracy

If however the receiver can receive a fourth satellite, it would find that the fourth circle would only intersect at the same point as the others if the receiver applied an adjustment to its own clock of a particular value. It can therefore correct its clock to a very high degree of accuracy - so much so that your GPS clock is probably the most accurate clock you have (when it is receiving satellite signals).

Fact: The minimum number of satellites for a "2D" position fix on the ground is three satellites. The accuracy gets much better if you have at least four.

What is GPS and How does it Work?   **GPS Inaccuracy**

Altitude - Positions in "3D"

So far circles have been considered, as if the Earth is flat. Of course it isn't and in addition aircraft operate in three dimensions. Actually the signals from the satellite radiate out in a spherical shape. If it takes a minimum of three satellites to find your position in "2D" space you can easily see how adding a fourth can find your position in "3D" space:

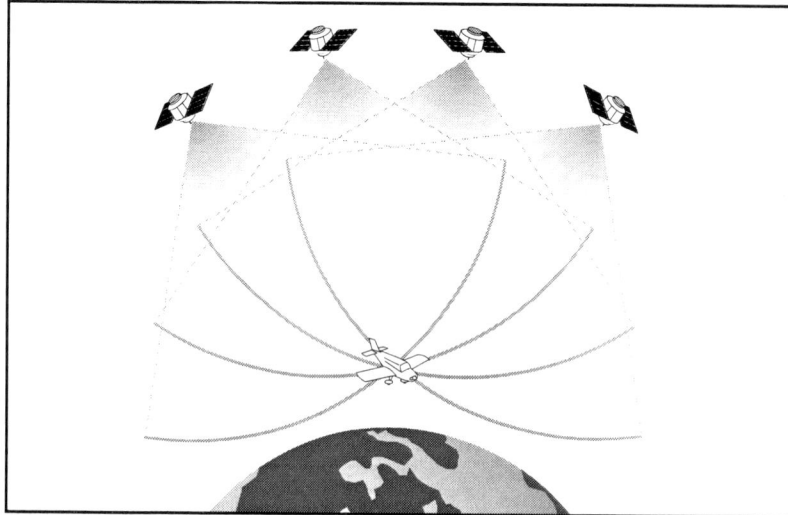

Position in three dimensions

Of course you still have the receiver's clock error to contend with so you will get a much better fix in "3D" space if you have a fifth satellite to enable the compensation for the receiver clock accuracy.

Fact: The minimum number of satellites for a "3D" position fix including altitude is four satellites. The accuracy gets much better if you have at least five.

This gives a point in space that has to be related to the surface of the earth. Most GPS receivers now have a fairly good model (called a "geodetic" model) that does this and takes into account to a certain degree the fact that the earth is not truly spherical, and computes this into an altitude display.

Accuracy

Accuracy of a reported position on the Earth's surface is +/- 15 metres using typical modern equipment, though frequently may be better. In general, the more satellites that can be received the more accurate the position information available, and although you need just four there are often at least eight or more satellites "visible" above the horizon.

Generally speaking, the accuracy of a GPS's "3D" positional accuracy is about one and a half times worse than the horizontal accuracy. Before "selective availability" (see below) was turned off the margin for error was typically at least 450 feet. Now the margin for error is closer to 75 feet, not taking into account errors in the geodetic model.

> **⚠ Alert:** You should never normally use your GPS altitude display as your primary height reference, especially not to determine ground separation.

Error Factors

There are a number of factors affecting the accuracy of the position indicated by the receiver:

- The average speed of the signal is affected by changes in the layers of the atmosphere, especially the ionosphere and troposphere.
- The exact satellite positions can vary slightly and even though they are constantly monitored and adjustments made there can still be errors. These errors are called "ephemeris" errors.
- Satellite Geometry. The wider the angles between the satellites the better. It is relatively easy to understand how this would create a clearer intersection between the spheres described above. If the visible satellites are "close together" the accuracy is degraded.
- The signal may bounce off objects and create "multi-path" errors. Most modern aviation GPS receivers are relatively immune to multi-path effects.
- Selective Availability (SA). Selective availability was the introduction of deliberate random errors by the US Department of Defense to avoid an accurate signal being used for missile targeting by enemy powers. Use of selective availability was discontinued in May 2000 (though could be reintroduced at any time should military circumstances warrant it).
- Deliberate jamming. Military powers can deliberately jam the signal rendering it useless in particular areas. The dates of tests usually appear in the UK AIC's.

New Developments

You may have heard of new developments in accuracy such as Differential GPS (DGPS) and Wide Area Augmentation System (WAAS). These are discussed in Appendix A. The GPSMAP 196 and 295 can support WAAS.

Familiarising yourself with the GPS

This Section looks at how you power the GPS, its various pages ("screens") of information and how to use its basic features. There is some ground school to make sure you are fully conversant with the basic working of the GPS before you install the GPS in the aircraft and take to the air.

There is a step-by-step guide to the main features of your GPS and a summary "map" showing how all the pages fit together.

This section tells you most of the basics and is intended to supplement rather than replace the manual that comes with your GPS. It's worth keeping your manual handy if you want more detail on a particular feature.

Introduction

This Chapter looks at the best ways to power the GPS.

All the Garmin systems considered in this book can take power from an appropriate number of AA batteries or from a cigarette lighter adapter (the latter being an option on some models). In addition the GPSMAP 195 has a removable battery pack that can be replaced by an optional rechargeable battery.

Batteries

The battery life varies dramatically between models and the kind of operations, especially factors like use of backlighting and whether the device is actually receiving satellites (you will find that batteries last significantly longer on the ground in simulation mode than in the air). You should use good alkaline batteries (or even the latest Lithium batteries which claim to last up to three times longer).

Tip: When you insert the batteries pay particular attention to the polarity. On some Garmin models some of the springs in the battery compartment need to be at the positive(+) end of the battery rather than the flat end, and the unit can sometimes work (albeit briefly) if some of the batteries are in the wrong way.

All of the models have a battery meter.

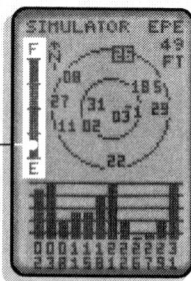

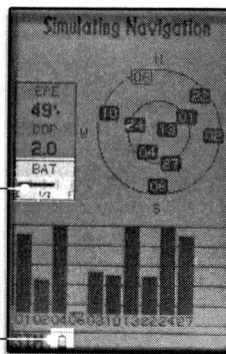

Battery Meter GPS 92 | Battery Meter GPS III Pilot | Battery Meter GPSMAP 195 | Battery Meter GPSMAP 295

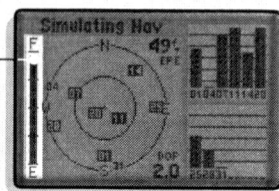

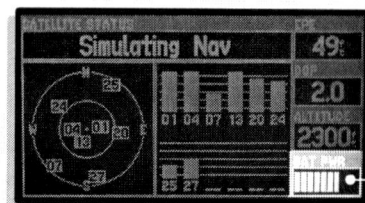

The typical battery life is quoted as 24 hours for the GPS 92, 8-10 hours for the GPS III Pilot (more on recent models), 8-10 hours if you have a GPSMAP 195 and only 2.5 hours if you have a GPSMAP 295.

Before starting a journey it is important to ensure that sufficient battery life remains. The battery meters tend to stay fairly full for a significant period of the battery life, especially on the GPSMAP 295. Experience has shown that for a full day's summer flying it pays to start the day with a fresh set of batteries in the GPS, and keep a spare set in reserve. If you have the GPSMAP 295, you will need three sets to ensure you have one set in reserve by the end of the day!

Airmanship: Don't start a flight unless the battery meter is reading at least half full on the GPS 92, GPS III Pilot and GPSMAP 195, or totally full on the GPSMAP 295.

Tip: The first few times you use your GPS, keep a note of how long the batteries last for, so you have a better understanding of the battery level indications.

Rechargeable Batteries

It is possible to use rechargeable AA batteries in the GPS (you can even use them in the GPSMAP 195 standard battery pack if you don't have the optional rechargeable pack). You will especially want to consider rechargeables if you have a GPSMAP 295 as it eats batteries for breakfast, lunch and tea!

Traditionally, rechargeable batteries have a much shorter life than alkaline batteries and they also have a different discharge characteristic and go from being apparently full to empty much more quickly. If you are using rechargeable batteries you should change the setting of the battery meter to make sure it reads correctly for the type of batteries you are using (how to do this is described in Chapter 8).

There are now two main types of rechargeable batteries that are readily available: Nickel Cadmium (NiCd or "NiCad"), and Nickel Metal Hydride (NiMH).

NiCads only store a relatively small charge and suffer from the "memory" effect if they aren't used properly and therefore aren't recommended (you may remember this from early mobile phones where the battery used to hold less and less charge until it became useless).

NiMH batteries are much better as they last about twice as long as Nicads, and whilst they still don't last as long as alkaline ones, disciplined use makes them a very good option. NiMH AA batteries typically store between 1200mAh and 1700mAh (compared with the 500mAh to 800mAh of NiCads), and it's worth hunting for the highest value you can find (try Tandy or Maplin Electronics). Of course, you will also need a battery charger designed for NiMH batteries!

NiMH batteries, like most other rechargeable types, will lose some charge while they are not being used. It is well worth checking the state of batteries if they haven't been used for a few weeks.

> **Tip:** You can "re-cycle" your rechargeable batteries the day before flight by leaving your GPS on to discharge them and then recharging them fully overnight. Keep a spare set of unopened Alkaline batteries as your reserve - they don't lose their charge quickly (though keep an eye on their "use by" date).

External Power

If the plane you are going to fly has a cigarette socket you can power the GPS directly from the plane's power supply. You will need a suitable power cable that is supplied with some models and optional with others.

> **Fact:** In some aircraft types (notably Cessnas) the cigarette lighter sockets may have been the subject of a mandatory disconnection in the past.

All of the Garmin GPS receivers are quoted as being very tolerant to the supply voltage and should work satisfactorily with both 12 volt and 24 volt supplies. On some receivers, when you are using external power the battery power meter doesn't appear and on some units is replaced by a little plug symbol.

When using the external power cable, be especially careful that it cannot tangle in the controls. Make sure that the power plug is fully inserted into the cigarette socket and isn't likely to work loose in flight.

> **Tip:** If you also have batteries in the unit, it will switch to battery mode rather than switch off if the cigarette plug works loose.

Unfortunately the Garmin GPS receivers don't charge rechargeable batteries even if they are plugged into an external supply.

Introduction

This Chapter steps through what happens when you switch your GPS on, the standard messages that you get and how the GPS acquires satellites to find its position.

You will want to do the first switching on with your feet on the ground! Ideally you will be outside or at least by a window, as one of the first things you will be doing is locating your current position.

You might want to read through this Chapter first before you work through it with the GPS.

> **Fact:** The GPS behaves slightly differently the first time it is switched on after it comes from the factory - you should take this into account if you have already used your GPS prior to getting this book!

Switching On

Make sure you have batteries correctly installed and have connected the antenna.

> Press and hold the red power button on your GPS briefly until the unit switches on.

The unit will start up and go through its power on sequence. This consists of a "title screen" while the unit conducts a self test.

Next the GPS will display information about the Jeppesen aviation database. The Jeppesen database includes all the airfields and navigation aids that the GPS knows about.

You need to make sure that the database is reasonably current and covers the area you are going to be flying in. This is covered in more detail in Chapter 8. For the time being you are probably more interested in getting up and running.

All the receivers except the GPS 92 will also display what Garmin call the "pilot's warning", notifying you that the GPS is intended as a supplementary aid to VFR navigation only. You should accept this warning by pressing the indicated key.

Alert: At the time of writing none of Garmin's portable units are rated for use as anything other than an auxiliary aid to navigation in VFR. Operating the GPS safely and getting the best out of it in accordance with this warning is what this book is about!

Start-up Sequence

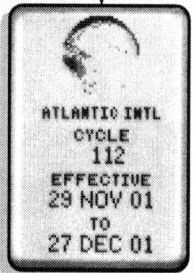

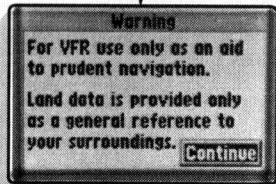

Note: If you have a GPSMAP 295 with MapSource cartridge installed then the additional screen (shown) will appear.

| GPS 92 | GPS III Pilot | GPSMAP 195 | GPSMAP 295 |

Next the GPS moves to the satellite acquisition screen. The receiver works in slightly different ways depending on the circumstances:

Autolocate mode: In this mode, the receiver behaves as if it knows nothing about its likely position. It starts "searching the sky" looking for satellites from scratch, updates its almanac, and will then start to "autolocate" itself. Depending on the model, it can take typically 2 to 5 minutes to locate itself (providing it has an aerial connected and has a good view of the sky). The receiver is normally factory set to enter this mode automatically when it is new. You can also manually put the receiver into this mode (see next page). When the receiver has found enough information it enters "Acquiring mode".

Acquiring mode: In this mode the receiver uses its last almanac and position information to acquire satellites. This typically takes less than 15 seconds if the receiver has just been used, for example on a previous flight (Garmin call this a warm start) or around 45 seconds if the receiver has not been used for a while but is near to the last location it was used in (cold start).

Autolocating & Acquiring

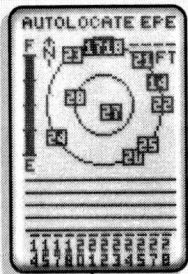

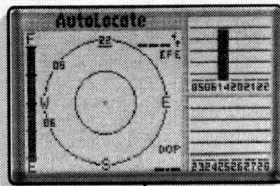

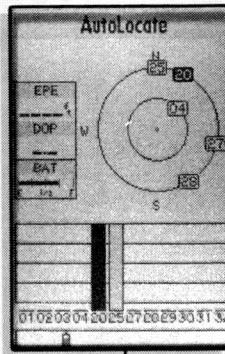

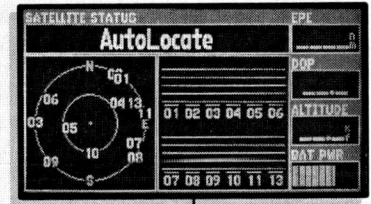

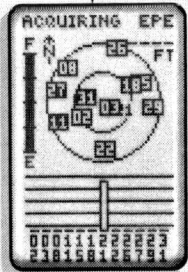

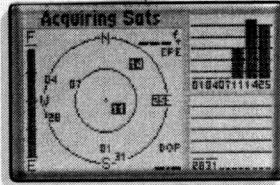

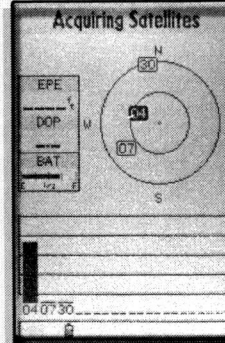

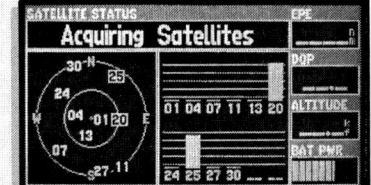

GPS 92

GPS III Pilot

GPSMAP 195

GPSMAP 295

Selecting an Initialisation Method

If the GPS has not been used before or if it has trouble acquiring any satellites (for example if you are using it indoors or if you haven't connected its antenna, or if you have moved several hundred miles since you last used it), it will prompt you to select an initialisation method.

- **Select Location.** You can tell the GPS roughly where you are to assist it to acquire the satellites more quickly. On the GPS III Pilot and the GPSMAP models you can do this by pointing at the map and zooming down until you identify roughly where you are. The GPS 92 doesn't have a "base map" (i.e. a map of land data including country outlines, roads, railways and rivers), so you need to tell it where you are from a list of countries.

- **Autolocate.** You can force the receiver into Autolocate mode so it searches the sky for satellites and starts to work out your position from scratch.

- **No Re-initialisation (Continue Acquiring).** The GPS will carry on trying to acquire satellites based on its last known information. You should check that the antenna is properly connected and that you are outside and the view of the sky isn't obscured by high buildings.

Selecting an Initialisation Method			
GPS 92	**GPS III Pilot**	**GPSMAP 195**	**GPSMAP 295**
The GPS will display a message "**Need to Select Init Method**".	The GPS will display a message "**Need to Select Init Method**".	The GPS will display a message "**Need to Select Init Method**".	The GPS will display a message such as "**Poor Satellite Reception**".
Next you should choose from the options:	Next you should choose from the options:	Next you should choose from the options:	Next you should choose from the options:
1. Select Country from List. This enables you to choose from a list of countries	**Use Map.** This enables you to scroll the map and zoom in and out to mark your location.	**Select from Map.** This enables you to scroll the map and zoom in and out to mark your location.	**Start Simulator.** This option anticipates you are using the GPS inside for training and enables you to enter simulator mode (see later).
2. Autolocate. This forces the GPS into Autolocate mode.	**Autolocate.** This forces the GPS into Autolocate mode.	**Autolocate.** This forces the GPS into Autolocate mode.	**New Location.** This lets you to choose between Automatic, which forces the GPS into Autolocate mode or Use Map, which enables you to scroll the map and zoom in and out to mark your location.
3. No Re-Init (Continue Acquiring). This forces the GPS to continue acquiring satellites.	**None.** This forces the GPS to continue acquiring satellites.	**No Re-Init.** This forces the GPS to continue acquiring satellites.	**Continue Acquiring.** This forces the GPS to continue acquiring satellites.

Selecting an Initialisation Method

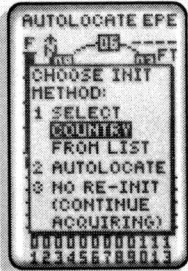

GPS 92

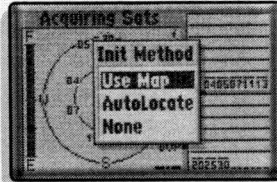

GPS III Pilot

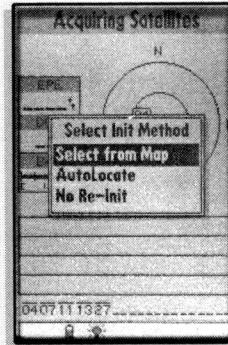

GPSMAP 195

GPSMAP 295

After acquiring the satellites the Satellite Status screen will continue to display information about the satellites received and their signal strengths and the type of navigation available. As you learned in Chapter 2, the minimum number of satellites for 2D navigation is three and for 3D navigation is four. In addition the receiver will display an Estimated Position Error (EPE) which is the accuracy that the position is believed to be, and a Dilution of Precision measurement (DOP) which is an indication of the satellite geometry. The lower this number the better.

After locating your position, the receiver will turn automatically to the map display page (GPS III Pilot and GPSMAP models) or the Position Page (GPS 92).

Initialising Position from the Map

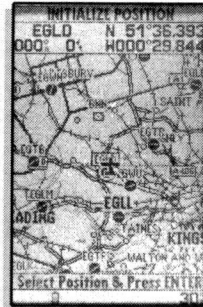

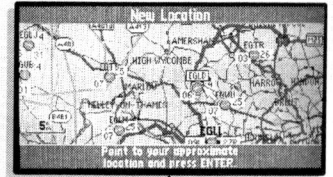

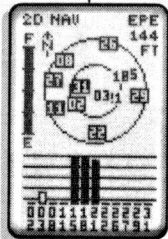

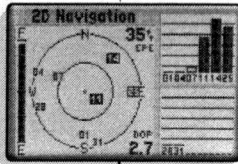

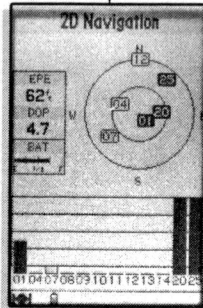

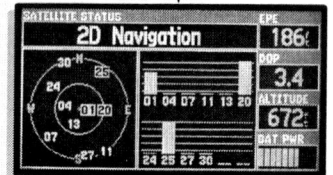

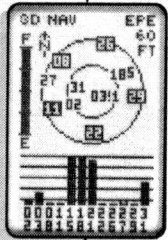

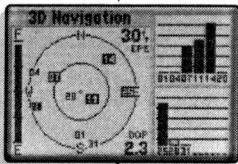

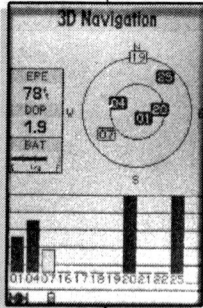

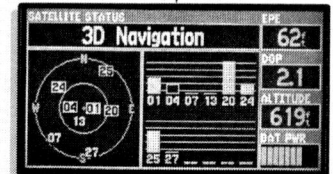

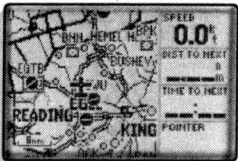

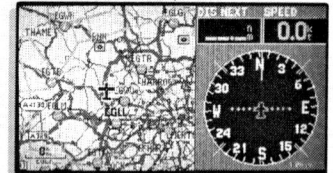

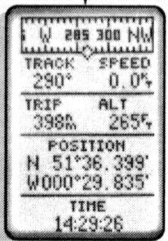

GPS 92

GPS III Pilot

GPSMAP 195

GPSMAP 295

Switching on the GPS

Initialising Position from the Map

Introduction

In this Chapter take time to become familiar with the GPS and its basic functionality. When you have finished this Chapter you will know about all the main pages or "screens" of your GPS.

Time spent on the ground is time well spent and if you get to grips with the way the user interface is put together in this Chapter you will have no trouble at all using the GPS in real time in the air.

A GPS is a complex piece of equipment packed into a very small space, and so it's hardly surprising that the "user interface" can be a little daunting at first sight.

> ⚠️ **Alert:** Work through this section on the ground. Don't be tempted to go airborne until you can find your way around the GPS without thinking twice about it.

Operating Principles

Understanding the logic behind the Garmin user interface makes it much easier to find your way around the GPS.

The features and functions are grouped into three main types:

- **The main page "loop"**
  - these are the main screens you will look at in the cockpit.
- **Functions accessed directly from buttons**
  - what happens when you press each button.
- **Functions accessed from the Main Menu**
  - these are functions needed in configuring the GPS or in ground preparation, but rarely in flight.

This Chapter looks at the main pages. The next two Chapters look at the button functions and the Main Menu functions.

Navigating the Main Page "loop"

The main functions of the GPS are visible through its main pages ("screens") of information. If you have configured the GPS properly before flight it is possible to operate the GPS almost solely by accessing these pages.

On most of these pages are pieces of information containing information such as speeds, bearings, distances and so on. These areas of information are known as "fields" and, as you will see later, they can be configured to provide you with almost any information you need.

You can change the page displayed by pressing the (PAGE) button. The main pages are in a loop, such that if you keep on pressing (PAGE) to change the page displayed you soon loop back to where you started.

You can move backwards around the loop by pressing the (QUIT) button.

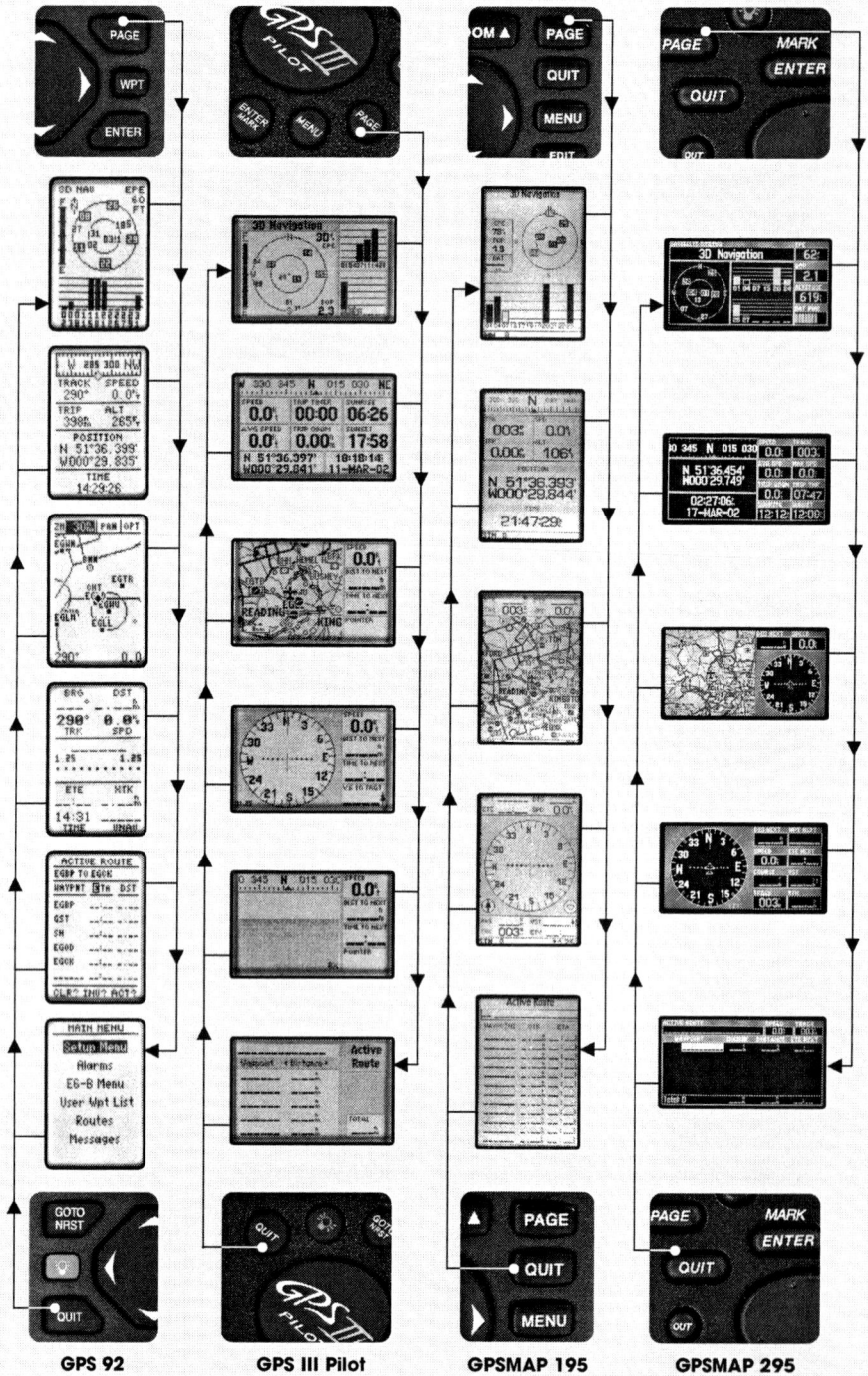

GPS 92   GPS III Pilot   GPSMAP 195   GPSMAP 295

Pages in the Main Page "Loop"

Here are the pages:

Satellite Status Page. This page displays the satellite status including which satellites are being received, how strong the signal, the almanac data (i.e. a picture of what satellites should be in view), and estimated position accuracy. It also contains the battery power meter. The GPSMAP 295 includes an altitude display.

Satellite Status Page

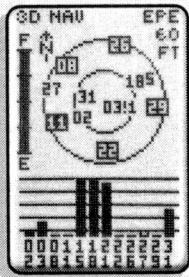

GPS 92

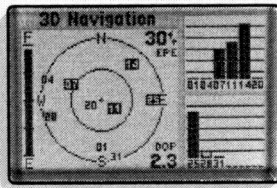

GPS III Pilot

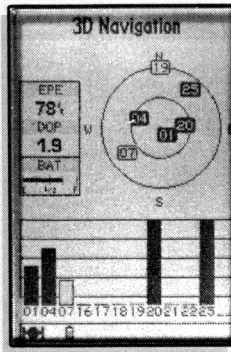

GPSMAP 195

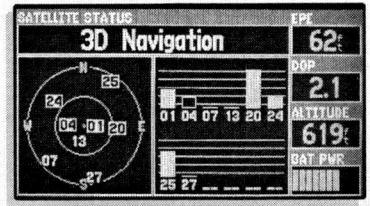

GPSMAP 295

Position Page: This page displays your latitude and longitude and has a compass segment that shows your magnetic heading. It also displays the time. Depending on the GPS model you can also see your track, speed, trip settings, altitude and sunrise and sunset times and you can configure these fields to show other information.

Position Page

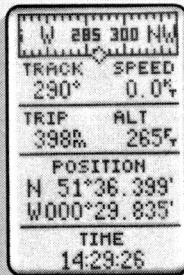

GPS 92

GPS III Pilot

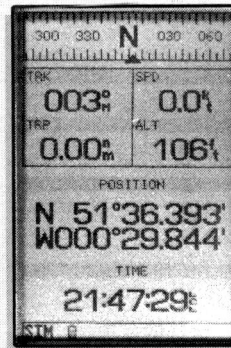

GPSMAP 195

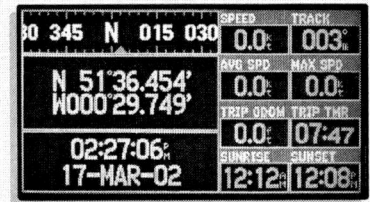

GPSMAP 295

Familiarising Yourself with Your GPS   **Satellite Status & Position Pages**

Moving Map page: This page is perhaps the most useful in-flight page and shows your position on a moving map as well as other useful items of data such as speed and distance to the next waypoint. With the exception of the GPS 92 the systems have a base map that has roads, railways, rivers, towns and other useful data - use the zoom in and out keys to adjust the scale. If you move left or right or up and down using the rocker pad the map will pan ("scroll") and a small display will show the position of the "cursor"(pointer or crosshair). By default the GPSMAP 295 also has an HSI on this screen. The data fields that are displayed by default vary significantly depending on the GPS model. The map page is configurable and it's well worth doing so for best results. You can configure both the map part of the display and the data fields displayed - this is described in Chapter 9.

Moving Map Page

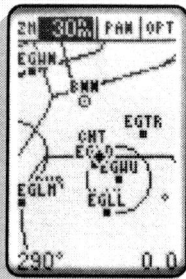 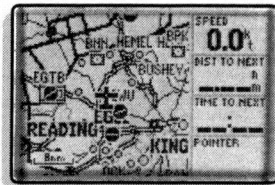  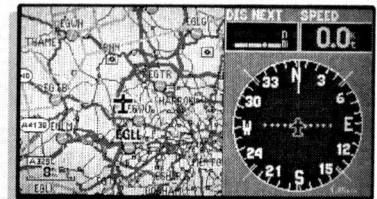

| GPS 92 | GPS III Pilot | GPSMAP 195 | GPSMAP 295 |

HSI page (not GPS 92). This page shows a representation of a Horizontal Situation Indicator, and various other useful data items such as speed and distance to the next waypoint. This page can be re-configured to display other data fields.

 **Fact:** The GPS Horizontal Situation Indicator is very similar to the HSI found in some aircraft. If you haven't come across an HSI before, it is as if a VOR course deviation indicator (CDI) has been superimposed over a Direction Indicator.

Navigation (CDI) page (GPS 92 only). Instead of the HSI page, the GPS 92 has a navigation page which has a display similar to a VOR Course Deviation Indicator (CDI) and other useful data items such as speed and distance to the next waypoint.

HSI Pages & CDI Page

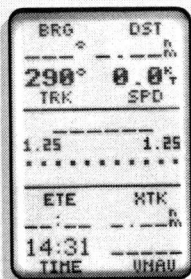

GPS 92

GPS III Pilot

GPSMAP 195

GPSMAP 295

Highway Page (GPS III Pilot only):

This page is unique to the GPS III Pilot and displays a "highway in the sky". It provides an interesting visualisation of the route. (It doesn't display much until a route is active).

Highway Page

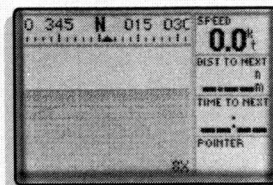

GPS III Pilot

Active Route Page: This page approximates to your paper flight log and is dynamically updated in real time. It shows the active route including information for each leg including the distance, time en-route and desired track (course) to the next waypoint. On the GPS III Pilot you need to use the left/right rocker pad to see the time en-route (and other information). This page only appears on the GPS 92 when a route is actually active, so you probably won't see it yet.

Active Route Page

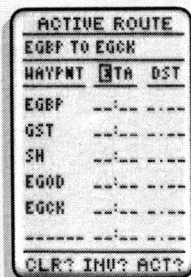

GPS 92

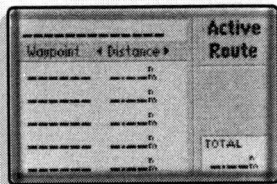

GPS III Pilot

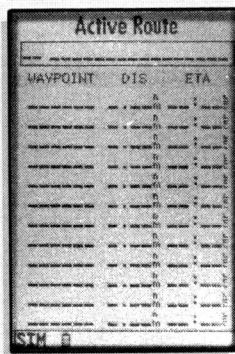

GPSMAP 195

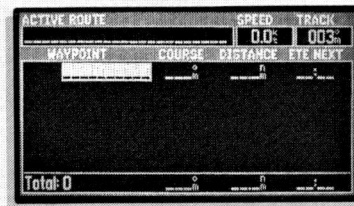

GPSMAP 295

Main Menu Page (GPS 92): Unlike the other models, the GPS 92 doesn't have a ⬤MENU button so its main menu page is accessed through the main page loop.

Main Menu Page

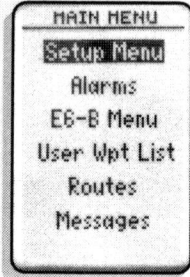

MAIN MENU
Setup Menu
Alarms
E6-B Menu
User Wpt List
Routes
Messages

GPS 92

Introduction

You've already met a few of the important buttons. This Chapter looks at them in detail, as well as all the other buttons you haven't used so far.

> **Alert:** Work through this section on the ground. Don't be tempted to go airborne until you can find your way around the buttons on your GPS without thinking twice about it.

GPS 92 Button Locations

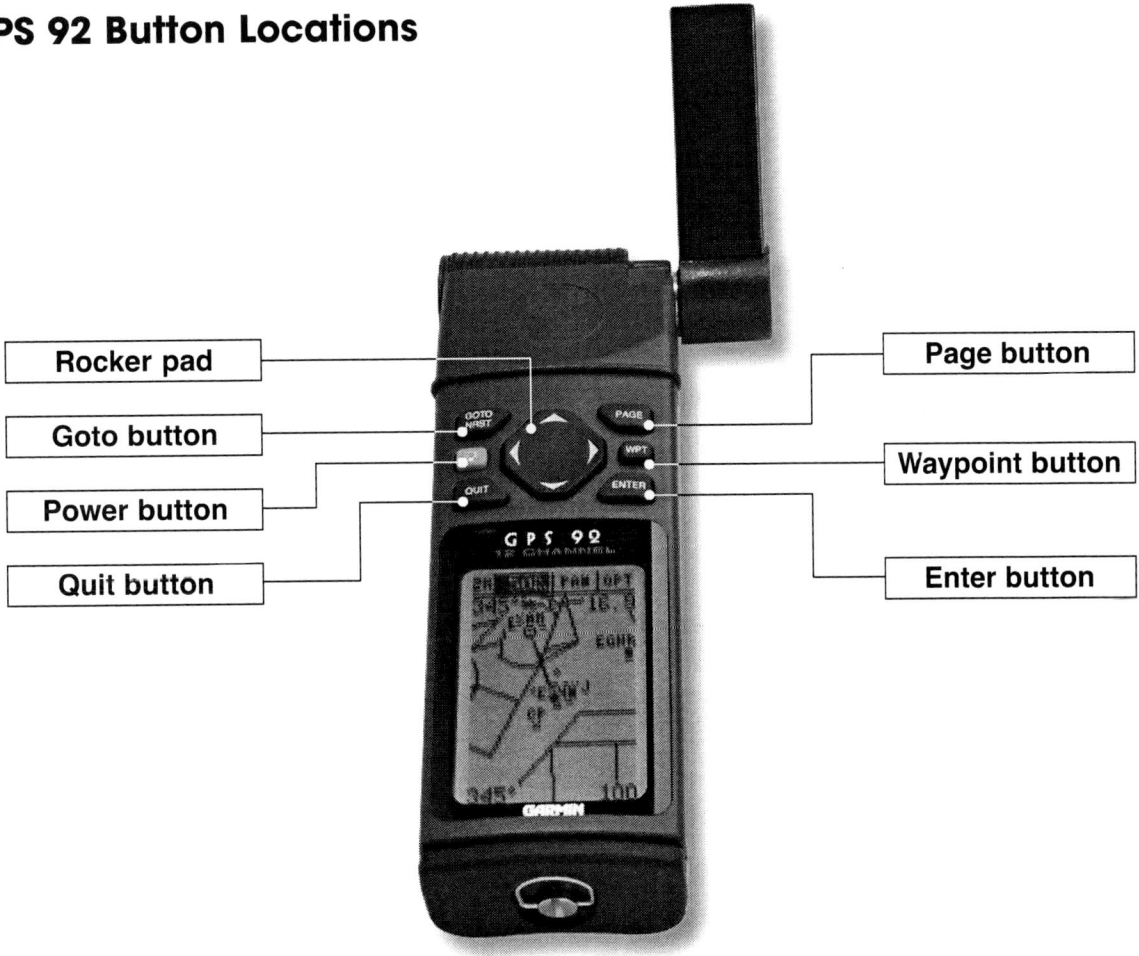

Rocker pad

Goto button

Power button

Quit button

Page button

Waypoint button

Enter button

GPS III Pilot Button Locations

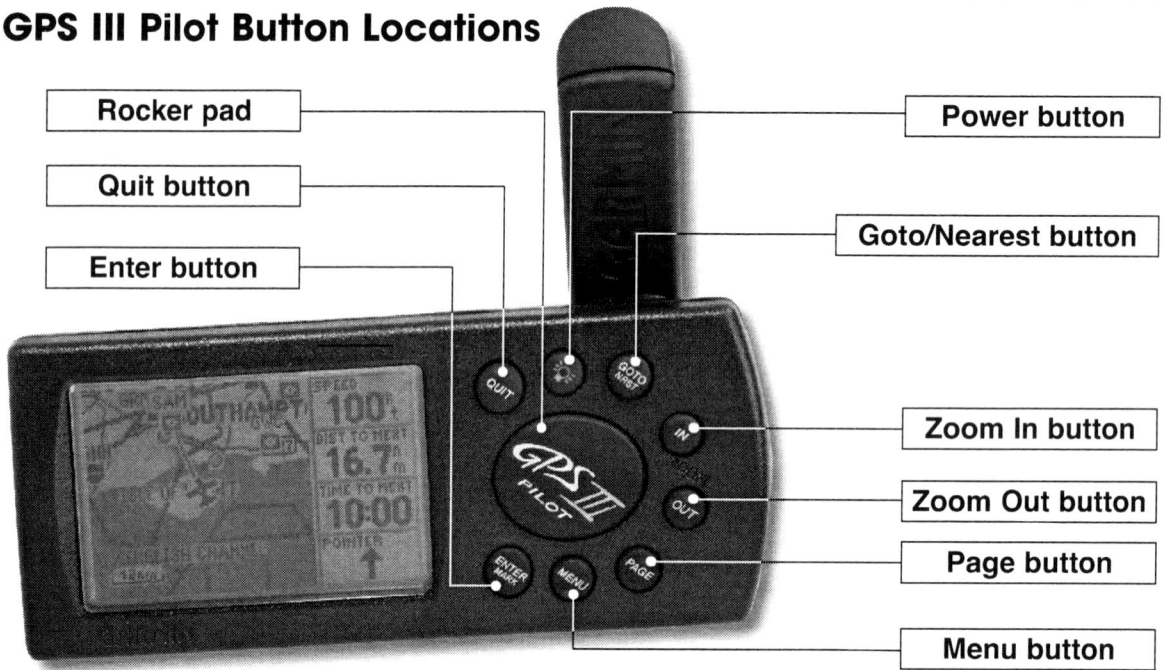

Rocker pad

Quit button

Enter button

Power button

Goto/Nearest button

Zoom In button

Zoom Out button

Page button

Menu button

GPSMAP 195 Button Locations

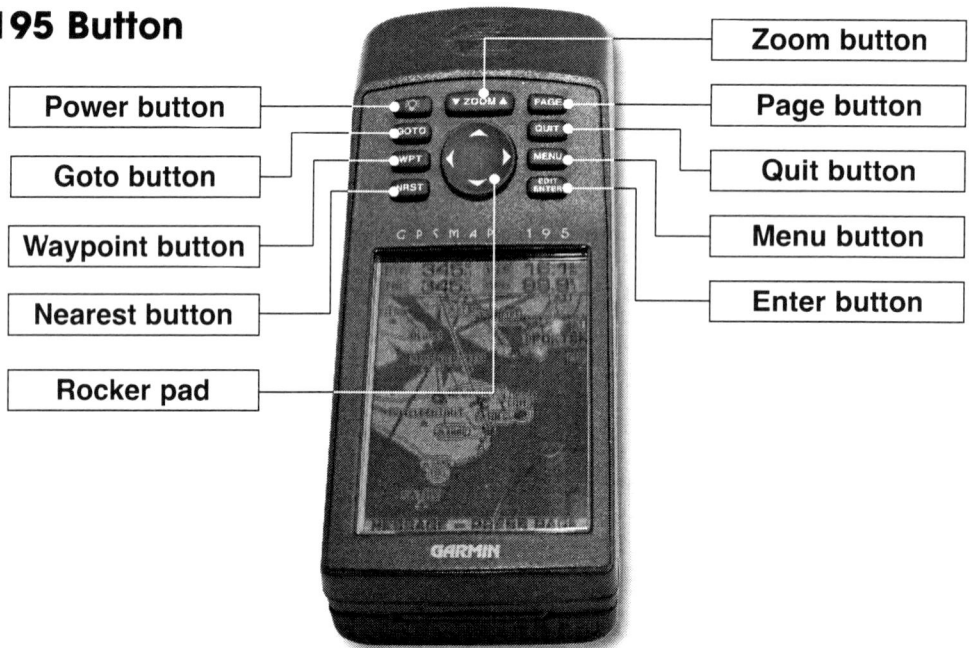

Power button

Goto button

Waypoint button

Nearest button

Rocker pad

Zoom button

Page button

Quit button

Menu button

Enter button

GPSMAP 295 Button Locations

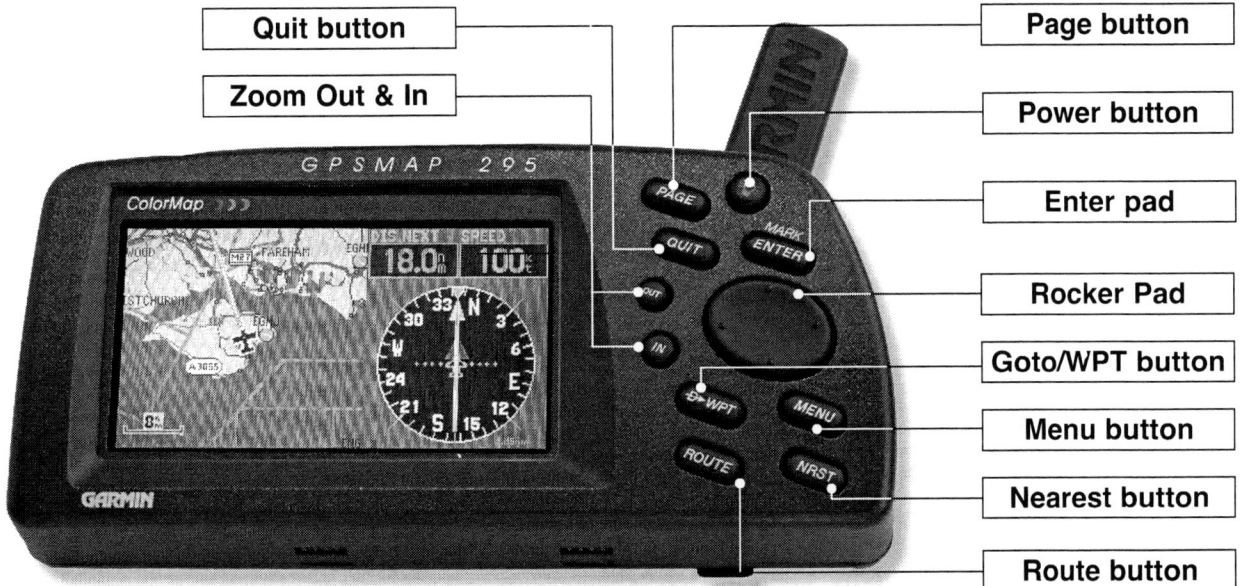

Quit button

Zoom Out & In

Page button

Power button

Enter pad

Rocker Pad

Goto/WPT button

Menu button

Nearest button

Route button

Button Functions

Let's take a look at the main functions of the buttons. For the most part all the models work in pretty much the same way, however you will notice that there are a few subtle differences here and there.

Tip: It's worth noting that some buttons have one function if pressed once and a second function if they are pressed twice in quick succession - remember this in turbulent conditions when you may involuntarily double-press a button.

Power button. The red Power button switches the unit on and off. Hold it down briefly to switch on. The unit will go through its power up sequence as described in Chapter 4. To turn the unit off hold the button down for about two seconds. Tapping the button briefly will enable you to set the backlight illumination to one of three settings (GPS 92, GPS III Pilot and GPSMAP 195), or set the contrast and brightness (GPSMAP 295). The backlight will normally turn off after a few seconds until you next press a button.

Rocker pad. The rocker pad enables you to carry out any function on the GPS that needs up/down or left/right movement or selection capability. You can use the rocker pad to scroll (pan) the map, move the cursor highlight on screens where it is present or scroll through characters when entering or editing waypoints. If you hold down the pad in one direction it will scroll more quickly.

Zoom buttons (not GPS 92). The zoom buttons enable you to zoom in and out on any screen displaying a map. They also act on HSI screens to change the CDI sensitivity, and the Highway scale (GPS III Pilot). The GPS 92 doesn't have zoom buttons - instead you set the scale of the map by editing the scale field on the Map page.

Page button. The ⬤PAGE button enables you to page forwards through the main pages as described in Chapter 5. It can also be used when prompted to proceed from certain pages back to the main pages.

Quit button. The ⬤QUIT button enables you to page backwards through the main pages as described above. It can also be used to cancel and exit or back out of most GPS options, including getting back to a normal operation if you have used the rocker pad to pan the map.

Enter button. The ⬤ENTER button is used to confirm a selection or entry. On the GPS III Pilot and GPSMAP 295 holding this button on a map screen will "mark" (set) the current position as a user waypoint. You will learn more about user waypoints in Chapter 15.

Menu button (Not GPS 92). The ⬤MENU button selects menus. If you press it once it opens up a context sensitive sub menu of commands relevant to the page you are on. If you press it twice in quick succession it takes you to the Main Menu page. On the GPS 92 the Main Menu page is one of the pages in the loop of main pages.

Goto button. This button gives you instant access to one of the most powerful and easy to use navigation features of the GPS. It enables you to create an instant route directly from your current position to a waypoint of your choice. The Goto feature is described in Chapter 19. The Goto button on the GPSMAP 295 looks like D with an arrow through it: ⬤D►WPT. On the GPS III Pilot it is combined with the "nearest" function (see below).

Nearest button. The ⬮NRST⬮ button displays a list of the nearest aerodromes and also enables you to access other feature types such as VORs, NDBs etc. The GPS 92 and GPS III Pilot combine the NRST functionality with the Goto button (on the GPS 92, press it twice and on the GPS III Pilot, hold the ⬮GOTO⬮ button down to access NRST), whilst the GPSMAP 195 and GPSMAP 295 have a dedicated NRST button. How to Goto the nearest aerodrome is described in Chapter 17 about Diversions.

Waypoint button. The ⬮WPT⬮ button enables you to find out additional information about a waypoint. On the GPSMAP 295 this function is combined with the Goto functionality on the ⬮Đ▸WPT⬮ button. On the GPS 92 and GPSMAP 195 pressing the ⬮WPT⬮ button twice in quick succession marks (sets) the current position as a waypoint (similar to the Enter/Mark facility on the GPS III Pilot and GPSMAP 295). There is no ⬮WPT⬮ button on the GPS III Pilot and so waypoint information is accessed through a menu (see next Chapter).

Route button (GPSMAP 295 only). The ⬮ROUTE⬮ button enables you to jump to the route set up page, which is where you can edit new routes (don't confuse this with the main active route page which looks similar). Only the GPSMAP 295 has a dedicated button for this. The other models access the routes page from the Main Menu.

Introduction

This Chapter introduces the GPS Main Menu. The Main Menu enables you to access many features involved in configuring the GPS on the ground as well as features you may need occasionally in the air.

The operation of the Main Menu varies between different models so each model is covered in turn.

The GPS 92 Main Menu

The GPS 92's Main Menu is part of the main page loop and enables you to access a number of sub-menus. You can select the sub-menu you need by using up/down on the rocker pad and pressing (ENTER).

Setup Menu. This sub-menu contains several sub-features including the Simulator, Units, position format and geodetic datum, audio and display, Time Zone, data interface and Nearest Airport settings.

Alarms. This sub-menu enables you to switch airspace alarms on or off, and also arrival and track errors. This page only affects airspace warning messages, not the display of airspace on the map.

E6-B. This sub-menu contains an electronic version of features that can be calculated using an E6B circular slide rule (American version similar to the Pooley's CRP-1 "wizzwheel"), including true airspeed and density altitude, and winds aloft. You can also compute trip and fuel settings, lookup sunrise and sunset times and configure vertical navigation settings.

User Waypoint List. You can enter up to 500 user waypoints, which can include towns, VRP's or any other waypoints you might find useful.

Routes. This enables you to create up to 20 routes of up to 30 waypoints each.

Messages. This displays any current warning or alerting messages.

GPS 92 Main Menu Tree

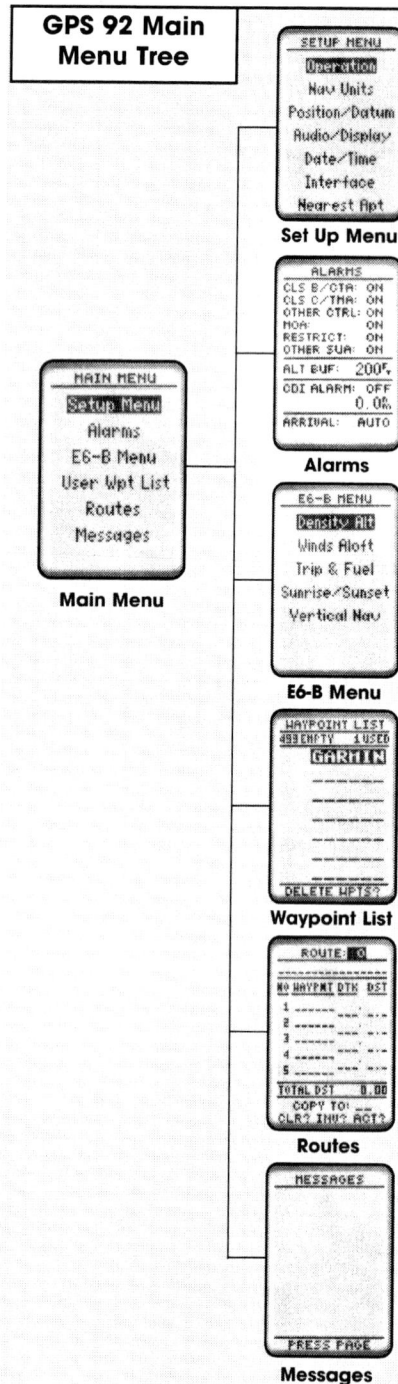

Set Up Menu

Alarms

E6-B Menu

Main Menu

Waypoint List

Routes

Messages

The GPS III Pilot Main Menu

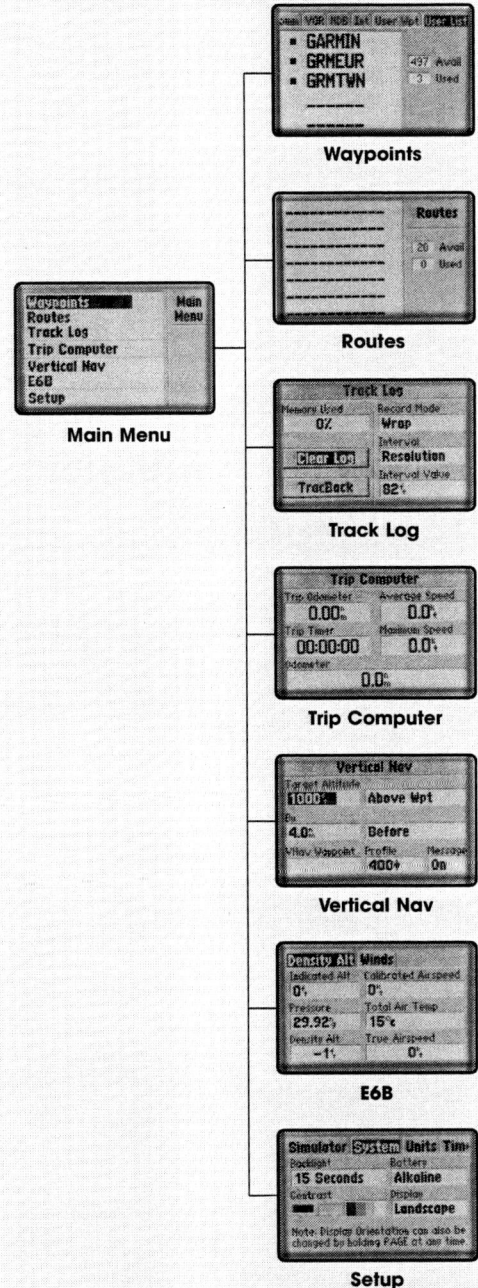

To access the GPS III Pilot's Main Menu, press the ⊙ Button twice in brief succession. The Main Menu enables you to access a number of sub-menus. You can select the sub-menu you need by using up/down on the rocker pad and pressing ⟨ENTER⟩. Some of the sub-menus themselves have further sub-menus which are displayed like file "tabs" similar to the representation found on computers and they can be selected using left/right on the rocker pad.

Waypoints. This enables you to see information about all the waypoints stored in the GPS including Airports, VORs, NDBs, Intersections and User waypoints.

Routes. This enables you to create up to 20 routes of up to 30 waypoints each.

Track Log. This enables you to configure the track log settings. The track log enables you to record your track. It will be displayed on the map screen as a dotted line.

Trip Computer. The trip computer records the distance, time and speed information for your flights.

Vertical Nav. This enables you to configure vertical navigation information and is typically used to let you know when to start your descent to an airfield.

E6B. This menu contains a couple of electronic features that can be calculated using an E6B circular slide rule (American version similar to the Pooley's CRP-1 "wizzwheel"), including true airspeed and density altitude, and winds aloft.

Setup. This menu contains numerous sub-features including configuring the Simulator, basic System settings, Units, Timers, Time Zone, Alarms, Airspace warnings, nearest airport criteria, position format and geodetic datum, and data interface.

Waypoints

Routes

Main Menu

Track Log

Trip Computer

Vertical Nav

E6B

Setup

The GPSMAP 195 Main Menu

To access the GPSMAP 195's Main Menu, press the (MENU) button twice in brief succession. The Main Menu enables you to access a number of sub-menus. You can select the sub-menu you need by using up/down on the rocker pad and pressing (EDIT ENTER).

Setup Menu. This sub-menu contains several sub-features including the Simulator, Airspace Alarms, Units, position format and geodetic datum, Nearest Airport Criteria, audio and display settings, Time Zone and data interface.

Routes. This enables you to create up to 20 routes of up to 30 waypoints each.

Track. This enables you to configure the track log settings. The track log enables you to record your track. It will be displayed on the map screen as a dotted line.

Timers. This enables you to use a manual timer and see other system timers including battery usage and trip timers.

Vertical Navigation. This enables you to configure vertical navigation information and is typically used to let you know when to start your descent to an airfield.

Density Alt/Winds Aloft. This sub-menu contains an electronic version of features that can be calculated using an E6B circular slide rule (American version similar to the Pooley's CRP-1 "wizzwheel"), including true airspeed and density altitude, and winds aloft.

Trip & Fuel Planning. You can compute trip and fuel settings, lookup sunrise and sunset times.

Weight & Balance. This particularly useful function is unique to the GPSMAP 195 and let's you quickly set up and compute a weight and balance schedule for the aircraft you will fly.

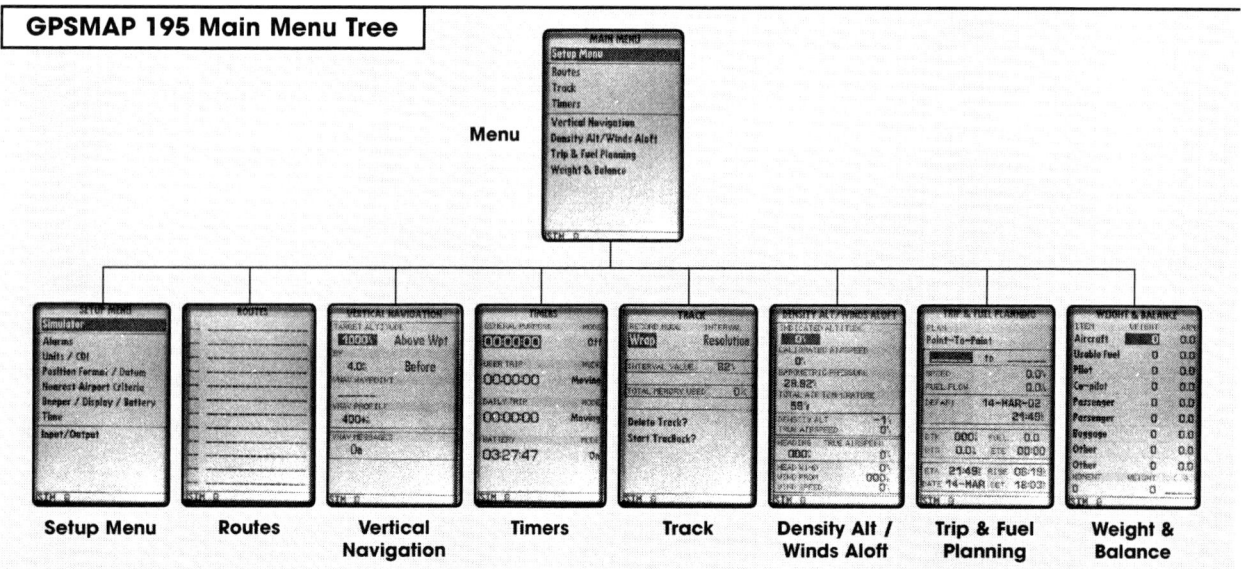

GPSMAP 195 Main Menu Tree

Menu

Setup Menu | Routes | Vertical Navigation | Timers | Track | Density Alt / Winds Aloft | Trip & Fuel Planning | Weight & Balance

The GPSMAP 295 Main Menu

To access the GPSMAP 295's Main Menu, press the ⬤MENU button twice in quick succession. The Main Menu enables you to access a number of sub menus displayed as file "tabs". The sub-menu tabs can be selected using left/right on the rocker pad.

System. This tab enables you to set the Usage Mode (Aviation or Land), Time Zone, Beeper, Battery and (later software only) WAAS setting. This screen also displays the operating software version.

Units. This tab enables you to set the units of measurement used by the GPS and also the position format and geodetic datum.

Timers. This tab enables you to use a manual timer and see other system timers including battery usage and time since midnight.

VNav. This tab enables you to configure vertical navigation information and is typically used to let you know when to start your descent to an airfield.

Track Log. The track log tab enables you to record your track. It will be displayed on the map screen as a dotted line.

Airspace. This tab enables you to switch airspace alarms on or off. This page only affects airspace warning messages, not the display of airspace on the map.

Alarms. This tab enables you to configure approach and arrival notifications, and off-course alarms. The system also has a clock alarm.

E6B. This tab contains an electronic version of features that can be calculated using an E6B circular slide rule (American version similar to the Pooley's CRP-1 "wizzwheel"), including true airspeed and density altitude, and winds aloft.

Interface. This tab enables you to configure the unit's data interface.

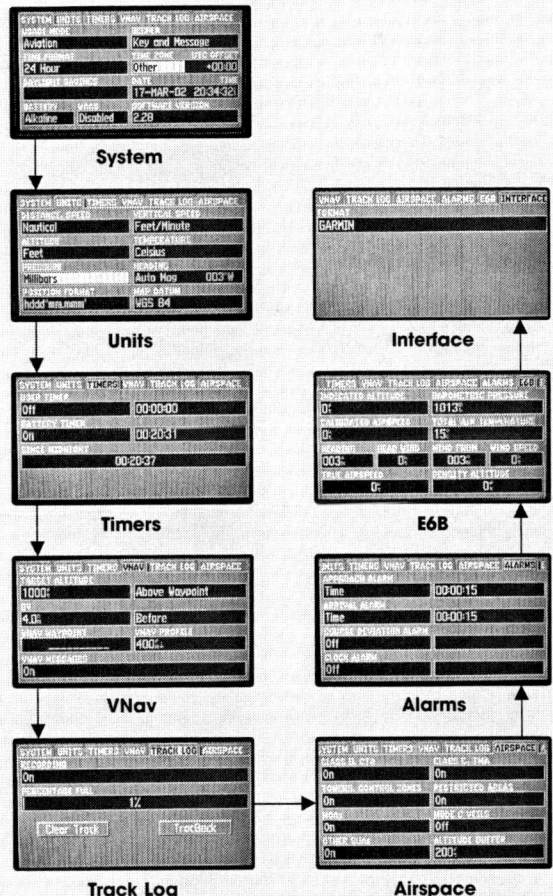

GPSMAP 295 Main Menu Tabs

System

Units   Interface

Timers   E6B

VNav   Alarms

Track Log   Airspace

Introduction

This Chapter shows you how to do some important configuration checks and make changes to the way your GPS is set up.

GPS is inherently an American system and Garmin is an American company and so it's not surprising that some of the standard "default" settings of your GPS are better suited to use in America. That means that you will have to do a certain amount of pre-configuration.

Alert: Work through this section on the ground. You don't want to be making fundamental configuration changes when airborne.

Fact: As you work through this section remember to use the ⬛ENTER⬛ button to activate selections and select a data "field" to edit its setting. Press ⬛ENTER⬛ again when you have finished editing the setting. If you make a mistake, press ⬛QUIT⬛ to back out.

Checking the Database

Switch your GPS on. During its power on sequence it will display the details of the Jeppesen database. On the assumption you are intending to use the GPS in Europe you will be looking to make sure the database is a recent Atlantic International version (Atlantic Intl).

Fact: Jeppesen make three databases for Garmin Units: Americas, Atlantic International and Pacific International.

If the database isn't Atlantic International (for example if you have imported the unit from the USA), you will need to update the aviation database. This is beyond the scope of this Chapter and is described in Chapter 22.

Next check the database is reasonably current. If the database is out of date you should consider updating it as described in Chapter 22.

Tip: If you have just bought your GPS and the database is more than three months old you may wish to try your luck with the retailer which sold you the unit and suggest they might throw in a database update certificate (again see Chapter 22). Most new GPS units now come with a free update key to enable you to do one free database update online.

Finally check the base map (except GPS 92). The base map contains all the non-aviation data such as roads, railways, rivers and towns. You are looking for a database that is applicable to Europe (e.g. International or Atlantic Highway).

Tip: Be careful if you imported the receiver, as the base map detail may not cover your area.

Checking the Database

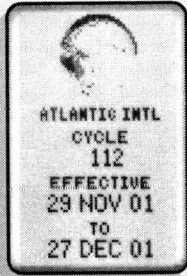

GPS 92

GPS III Pilot

GPSMAP195

GPSMAP 295

Important Configuration Checks — Checking the Database

Setting the Time Zone

The time zone can be set as well as the time format. In addition the time zone can be adjusted for daylight saving time if required.

> **Time Units:** This book uses UTC as the standard time base. You may prefer to set the local time to take account of British Summer Time or Central European Time for example.

> **Reminder:** To get to the Main Menu, press the (MENU) button in quick succession, except on the GPSMAP 92 where the Main Menu is in the main page loop. Select the option you need using the rocker pad.

Setting the Time Zone			
GPS 92	**GPS III Pilot**	**GPSMAP 195**	**GPSMAP 295**
From the Main Menu, select **Setup Menu** and then **Date/Time**. Ensure that the **Local Offset** is set to **+00:00**.	From the Main Menu, select **Setup** and then the **Time** tab. Select the **Time Format** and set it to **UTC**. Ensure the **Local Time Zone** is set to **00:00**.	From the Main Menu, select **Setup Menu** and then **Time**. Ensure the **Time Format** is set to **Local 24Hr.*** Ensure the **Local Time Zone** is set to **00:00 (Behind UTC)**. * Newer units also offer **UTC** as an option in which case it should be selected instead.	From the Main Menu, ensure the **System** tab is selected. Set the **Time Format** to **24 Hour**, the **Time Zone** to **Other** and the **UTC Offset** to **+00:00** (the **Daylight Savings** setting is not available when the Time Zone is set to Other).

Setting the Time Zone

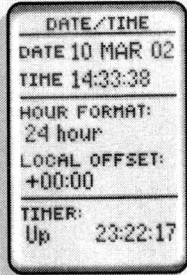
GPS 92

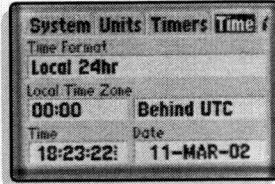
GPS III Pilot

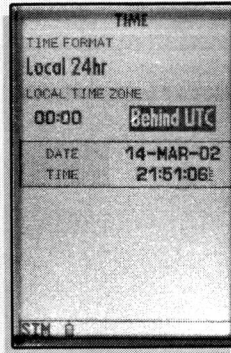
GPSMAP 195

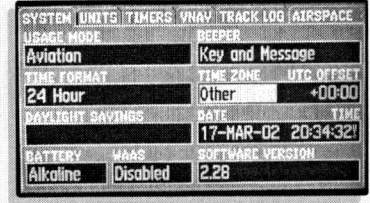
GPSMAP 295

Setting the Measurement Units

It's useful if all the standard units of measurement are set to the UK/European standards that you are used to.

Setting the Measurement Units			
GPS 92	**GPS III Pilot**	**GPSMAP 195**	**GPSMAP 295**
From the Main Menu, select **Setup Menu** and then **Nav Units.** Check that the **Dist/Speed** is set to **NM/Knots** and **Altitude** in Feet. You will probably need to change the **Temperature** to **Celsius** and the **Baro Pressure** from Inches of Mercury to **Millibars** (HectoPascals). Check that the **Heading** is set to **Auto.** The **CDI** setting affects the CDI sensitivity on the main Navigation page.	From the Main Menu, select **Setup** and then the **Units** tab. Check that the **Distance, Speed** Is set to **Nautical** (for knots), **Heading** is set to Auto Mag, and **Altitude** in **Feet.** You will probably need to change **Pressure** from Inches to **Millibars** (HectoPascals). Check that the **Vertical Speed** is in **Feet/Minute** and the **Temperature** is in **Celsius.**	From the Main Menu, select **Setup Menu** and then **Units/CDI.** Check that the **Distance & Speed** is set to **Nautical(nm,kt)** and **Altitude** in Feet. Check that the **Heading** is set to **Auto Mag Var.** Check that the **Vertical Speed** is in **Feet/Minute.** You will probably need to change the **Pressure** from Inches of Mercury to **Millibars** (HectoPascals) and the **Temperature** to **Celsius.** The **CDI** setting affects the CDI sensitivity on the main HSI page.	From the Main Menu, select the **Units** tab. Check that the **Distance, Speed** is set to **Nautical** (for knots), **Vertical Speed** is in **Feet/Minute, Altitude** in **Feet**, and the **Temperature** is in **Celsius.** You will probably need to change **Pressure** from Inches to **Millibars** (HectoPascals). Check that the **Heading** is set to **Auto Mag.**

Important Configuration Checks **Setting the Measurements Units**

Setting the Measurement Units

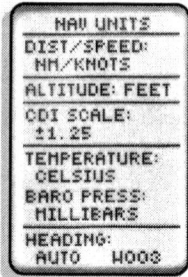

GPS 92

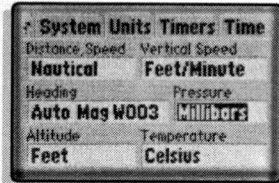

GPS III Pilot

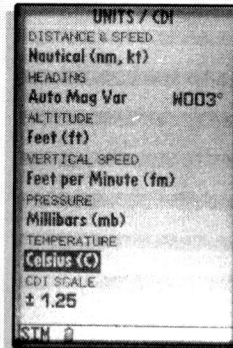

GPSMAP 195

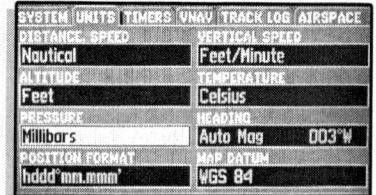

GPSMAP 295

Position Format & Map Datum

You need to make sure that the Position Format and Map Datum matches the other sources of information you have. As you learned in Section 1, there are a number of "geodetic models" that have been used to model the earth. A full discussion of geodetic models is beyond the scope of this book, however, in brief, there are very slight differences in the grid layout between models, due to the historical measurement. This means that a position quoted in one model may not quite map onto the same position in another model. You may remember news stories prior to the Millennium debating the "repositioning" of the Greenwich meridian - this was essentially a debate about the map datum used (though it wasn't reported as such).

Effective from 1st January 1998 the International Civil Aviation Organisation (ICAO) mandated the use of a geodetic model from the World Geodetic Survey 1984, called WGS84 for short.

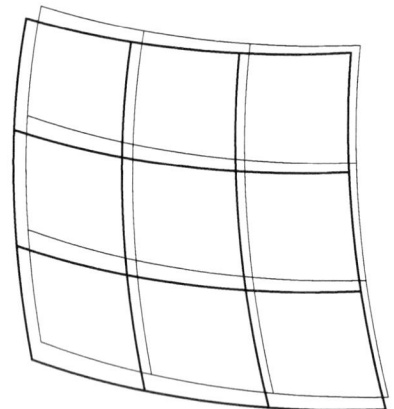

Grids/different Map Datums may not quite map onto each other.

Important Configuration Checks　　**Position Format & Map Datum**

Tip: There is a good web site dedicated to this subject at www.wgs84.com

Along with the WGS84 Map Datum, latitudes and longitudes are now generally expressed in degrees, minutes and decimal minutes (before this in the UK, they were generally expressed in degrees, minutes, seconds and decimal seconds).

Reminder: There are sixty minutes in a degree and sixty seconds in a minute. The earth is split into 360 degrees of longitude (180 East and 180 West) and 180 degrees of latitude (90 North and 90 South).

Alert: In practice, using the wrong map datum leads to errors ranging from a few feet to a few hundred yards.

Position Format & Map Datum			
GPS 92	**GPS III Pilot**	**GPSMAP 195**	**GPSMAP 295**
From the Main Menu, select **Setup** and then **Position/Datum.** Check that the **Position Format** is set to **hddd°mm.mmm'** (degrees and decimal minutes). Check that the **Map Datum** is set to **WGS 84.**	From the Main Menu, select **Setup** and then the **Position** tab. Check that the **Position Format** is set to **hddd°mm.mmm'** (degrees and decimal minutes). Check that the **Map Datum** is set to **WGS 84.**	From the Main Menu, select **Setup Menu** and then **Position Format / Datum.** Check that the **Position Format** is set to **hddd°mm.mmm'** (degrees and decimal minutes). Check that the **Map Datum** is set to **WGS 84.**	From the Main Menu, select the **Units** tab. Check that the **Position Format** is set to **hddd°mm.mmm'** (degrees and decimal minutes). Check that the **Map Datum** is set to **WGS 84.**

Important Configuration Checks   **Position Format & Map Datum**

Position Format & Map Datum

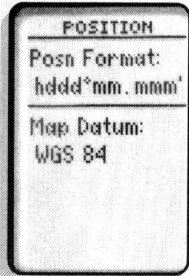

GPS 92

GPS III Pilot

GPSMAP 195

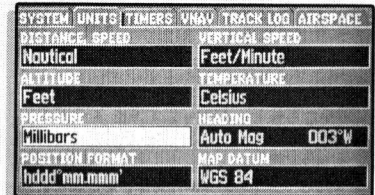

GPSMAP 295

Setting the Battery Type

Powering your GPS using different battery types was discussed in Chapter 3.

It is important that where possible you set the battery type correctly for the type of batteries you are using to make sure the battery meter reads correctly, since rechargeable and Alkaline batteries have different discharge characteristics.

Setting the Battery Type			
GPS 92	**GPS III Pilot**	**GPSMAP 195**	**GPSMAP 295**
The GPS 92 doesn't support a battery meter adjustment.	From the Main Menu, select **Setup** and then the **System** tab. In the **Battery** field select **Alkaline** or **Lithium** if you are using those types, **NiCad** if you are using rechargeables.	From the Main Menu, select **Setup Menu** and then **Beeper / Display / Battery.** In the **Battery Type** field select **Alkaline** if you are using Alkaline or Lithium batteries, **NiCad** if you are using rechargeables.	From the Main Menu, select the **System** tab. In the **Battery** field select **Alkaline** if you are using Alkaline or Lithium batteries, **NiCad** if you are using rechargeables.

Contrast & Brightness

You may wish to change the contrast or brightness/backlighting of the screen, though the default settings are normally OK for most purposes:

Setting the Contrast and Backlight			
GPS 92	**GPS III Pilot**	**GPSMAP 195**	**GPSMAP 295**
From the Main Menu, select **Setup Menu** and then **Audio/Display.**	From the Main Menu, select **Setup** and then the **System** tab.	From the Main Menu, select **Setup Menu** and then **Beeper/Display/Battery.**	Tap the [Power] button briefly.
You can now change the contrast control and the length of time the backlight will stay on after a button press.	You can now change the contrast control and the length of time the backlight will stay on after a button press.	You can now change the contrast control and the length of time the backlight will stay on after a button press.	You can now change the brightness and contrast using up/down and left/right on the rocker pad.
To Activate the backlight, tap the [Power] button briefly, once for dim lighting, twice for medium lighting and three times for high lighting.	To Activate the backlight, tap the [Power] button briefly, once for dim lighting, twice for medium lighting and three times for high lighting.	To Activate the backlight, tap the [Power] button briefly, once for dim lighting, twice for medium lighting and three times for high lighting.	The GPSMAP 295 display is always luminescent and doesn't have a back light, however it does have a day and night setting:
Tap a fourth time to turn it off.	Tap a fourth time to turn it off.	Tap a fourth time to turn it off.	From the Map page press [Menu] to bring up the context sensitive menu and select **Setup Map.**
Now when you press any key it will first turn on the backlighting for you.	Now when you press any key it will first turn on the backlighting for you.	Now when you press any key it will first turn on the backlighting for you.	On the **Map** tab you can change the colour mode between **Day, Night** and **Auto.**
			The dominant affect is that during the day time the map background is yellow, and at night it is black.
			If it is set to **Auto** it automatically changes mode at sunset and sunrise.

Tip: If you find you need to adjust the GPS 92, GPS III Pilot or GPSMAP 195 contrast in flight, use up/down on the rocker pad with the Satellite Status page displayed to activate the adjustment control, and left/right to make the adjustment - then press ⟨ENTER⟩.

Alert: Selecting the backlight to be always on can significantly reduce battery life.

Adjusting the Beeper Settings

(Not GPS III Pilot)

The GPS will beep audibly when you do anything from pressing a key to when it alerts you in any one of numerous ways, except for the GPS III Pilot, which doesn't have a beeper.

The beeping can become annoying when on the ground and is inaudible over the engine noise when airborne in most light aircraft cockpits. For this reason many people want to switch it off or at least modify some of the settings:

Adjusting the Beeper			
GPS 92	**GPS III Pilot**	**GPSMAP 195**	**GPSMAP 295**
From the Main Menu, select **Setup Menu** and then **Audio/Display.** You can adjust the **Tones** setting to activate on **Messages and Keystrokes** (button presses), **Messages Only** or **None.**	The GPS III Pilot doesn't have a Beeper.	From the Main Menu, select **Setup Menu** and then **Beeper / Display / Battery.** You can adjust the **Beeper** setting to activate on **Key** (button) **Presses and Message, Message Only** or **None.**	From the Main Menu, select the **System** tab and then **Beeper.** You can adjust the **Beeper** setting to activate on **Key** (button) **Presses and Message, Message Only** or switch it **Off.**

Adjusting the Display Mode

(GPS III Pilot only)

The GPS III Pilot is unique in that it can work either horizontally or vertically.

To change the mode, From the **Main Menu**, select Setup and then the **System** tab.

You can now change the **Display** setting between **Portrait** (for vertical use) and **Landscape** (for horizontal use).

Tip: Try to use the GPS III Pilot in the same mode when you do the exercises in this book as you will in the air, as the relative positions of the buttons is different between the modes. If you switch between modes you may find yourself pressing the wrong buttons in the air.

As a short cut you can also switch between Portrait and Landscape mode by pressing and holding down the `PAGE` button.

Introduction

This Chapter is dedicated to one of the easiest to use and most powerful features of your GPS, the GOTO button.

You are going to learn a lot of concepts in this Chapter including one of the most important - how to select waypoints from the Jeppesen database stored in your GPS.

All the Garmin portable GPS receivers have a Simulation mode to help you experiment with its operation without the expense of being airborne. This Chapter introduces you to the Simulator as well as to using Goto. It's important to understand how to use the Simulator as it is used a lot later in this book.

As you work through it, this Chapter will also introduce you to a number of other useful GPS concepts.

This Chapter is pretty intensive. You may wish to read through it before following it through on the GPS. Feel free to repeat it a few times with different destinations until you are happy you are comfortable.

> ⚠ **Alert:** Work through this section on the ground. Don't be tempted to go airborne until you can find your way around the Goto feature without thinking twice about it.

Using the Simulator

On this exercise you will fly from Denham to Cranfield and then to Oxford using Goto.

It doesn't matter whether you are indoors for this or even if you don't have an antenna attached as Simulator mode doesn't need a satellite signal.

Turn on your GPS and step through the initial pages until the Satellite Status page is shown. Now put the GPS in Simulator mode as shown on the next page:

Starting the Simulator

GPS 92	GPS III Pilot	GPSMAP 195	GPSMAP 295
From the Main Menu, select **Setup Menu** and then **Operation.**	From the Main Menu, select **Setup** and then the **Simulator** tab.	From the Main Menu, select **Setup Menu** and then **Simulator.**	From the Satellite Status page, press the [Menu] button once to bring up the context sensitive Menu.
Set the **Current Mode** to **Simulator.**	Set the **Mode** to **Simulator** On.	Set the **Mode** to **Simulator On.**	Select the **Start Simulator** option and press [Enter].
Use [Page] or [Quit] to get to the main Position page which shows speed, track and altitude fields.	The speed, track and altitude fields will be displayed.	The speed, track and altitude fields will be displayed.	Press the [Menu] Button again and select the **Set Altitude** option and press [Enter].
The **Speed** should be 0.0Kt. If it isn't, set it to **0.0Kt.**	The **Speed** should be 0.0Kt. If it isn't, set it to **0.0Kt.**	The **Speed** should be 0.0Kt. If it isn't, set it to **0.0Kt.**	Set the altitude to, say, **2300ft.**
Set the **Altitude** to, say, **2300ft.**	**Track Control** should be set to **Auto Track.**	Set the **Altitude** to, say, **2300ft.**	
	Set the **Altitude** to, say, **2300ft.**	**Track Control** should be set to **Auto Track.**	
		Your position will be set to the last known position.	

If you've done everything correctly, the Satellite Status page will say **Simulating Nav** and display lots of satellites.

Simulating Nav

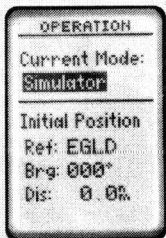
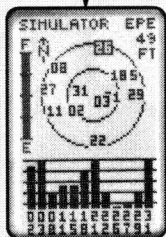
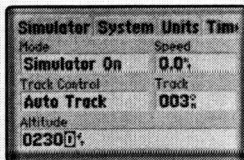
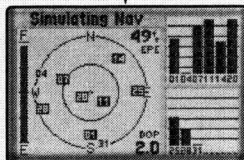

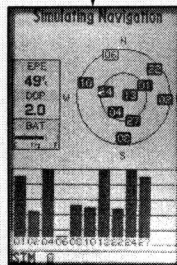
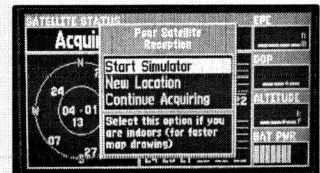
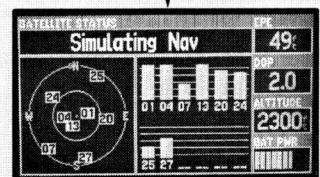

| GPS 92 | GPS III Pilot | GPSMAP 195 | GPSMAP 295 |

Tip: On the Garmin III Pilot and the GPSMAP 195, just as on the GPSMAP 295 you can also quickly enter Simulator mode by pressing the (MENU) button once on the Satellite Status page and selecting Start Simulator from the context sensitive menu.

Setting the Simulator Position

You can set the Simulator to any position you like. This exercise is going to start at Denham (EGLD), which is just North West of Heathrow.

Setting the Simulator Position			
GPS 92	**GPS III Pilot**	**GPSMAP 195**	**GPSMAP 295**
From the Main Menu select **Setup Menu** and then **Operation.** Set the Initial Position to EGLD (with a Bearing of 0 and Distance of 0).	From the Satellite Status screen press [Menu] once and select **Initialize Position.** Zoom in and highlight Denham (EGLD) north west of Heathrow (EGLL) and press [Enter].	From the Satellite Status screen press [Menu] once and select **Initialize Position.** Zoom in and highlight Denham (EGLD) north west of Heathrow (EGLL) and press [Enter].	From the Satellite Status screen press [Menu] once and select **Initialize Position.** Zoom in and highlight Denham (EGLD) north west of Heathrow (EGLL) and press [Enter].

Note: Denham is classified as a small airport. You will have to zoom in considerably before it appears. Later on you will learn how to make small airports appear when the map is zoomed further out.

Using Goto

First choose a waypoint to navigate to. In this exercise, the start point is Denham, and Cranfield (EGTC) will be your first destination:

Using Goto			
GPS 92	**GPS III Pilot**	**GPSMAP 195**	**GPSMAP 295**
Press the [Goto] button to get to the Goto page, and use the rocker pad to select the **Waypoint** field at the top of the screen.	Press the [Goto] button to get to the Goto page, and use the rocker pad to select the **Waypoint** field at the top of the screen.	Press the [Goto] button to get to the Goto/waypoints page, and use the rocker pad to select the **Waypoint** field at the top of the screen.	Press the [Goto] button to get to the Goto / waypoints page, and use the rocker pad to select the **Waypoint** field at the top of the screen.
Press [Enter] and start to edit the waypoint one letter at a time. Use the rocker pad up/down to select the correct letter, and the rocker pad left/ right to select which letter to edit.	Press [Enter] and start to edit the waypoint one letter at a time. Use the rocker pad up/down to select the correct letter, and the rocker pad left/right to select which letter to edit.	Press [Enter] and start to edit the waypoint one letter at a time. Use the rocker pad up/down to select the correct letter, and the rocker pad left/right to select which letter to edit.	Press [Enter] and start to edit the waypoint one letter at a time. Use the rocker pad up/down to select the correct letter, and the rocker pad left/right to select which letter to edit.
Edit the first four letters to read **EGTC** and press [Enter].	Edit the first four letters to read **EGTC** and press [Enter].	Edit the first four letters to read **EGTC** and press [Enter].	Edit the first four letters to read **EGTC** and press [Enter].
		Now make sure that **Goto** is highlighted on screen and press [Enter] again.	Now make sure that **Goto** is highlighted on screen and press [Enter] again.

Goto EGTC

GPS 92

GPS III Pilot

GPSMAP 195

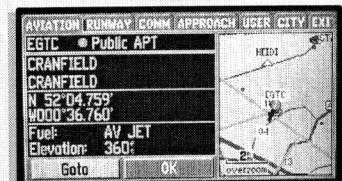

GPSMAP 295

"Goto" and Simulator Mode　　　　**Using Goto**

Taking Stock

At the moment your simulated plane is hanging in mid air 2000 feet above Denham. You can use the **PAGE** button to page through the different screens in the main page loop.

Take a particularly good look at the map screen. Try zooming it in and out. Remember to use the zoom buttons on the GPS III Pilot and GPSMAP 195 and GPSMAP 295 models, and edit the map scale field on the GPS 92 at the top left. On the GPS III Pilot GPSMAP 195 and the GPSMAP 295 there is a map scale at the bottom of the map.

If you zoom in too far the scale will be annotated with the word "overzoom" (not GPS 92). This means that you have zoomed closer than the accuracy of the base map provides for. On the GPSMAP 295 you can use the MapSource CD and a memory cartridge to provide greater detail, in which case the annotation will change to "mapsource" when the MapSource data is being used instead of the base map data.

> **Tip:** Different receivers use different map scale types. Map scales of 5nm to 8nm (GPS III Pilot and GPSMAP 295), and 20nm to 30nm (GPS 92 and GPSMAP 195) seem to be a good resolution to use when navigating en-route in the UK.

These example screens show each page in the main page loop of a GPS Pilot III shortly after activating Goto as described in this Chapter.

Setting the Simulator Speed

Now you need to simulate travelling to your waypoint:

Starting the Simulator Speed			
GPS 92	GPS III Pilot	GPSMAP 195	GPSMAP 295
From the main Navigation (CDI) page use the rocker pad to highlight the **SPD** field and press [Enter] to edit it. Select 100 knots.	From the main HSI page use the rocker pad up/down to increase/decrease the speed. The speed will be displayed in the **Speed** field. Select 100 knots.	From the main HSI page use the rocker pad up/down to increase/decrease the speed. The speed will be displayed in the **Speed** field. Select 100 knots.	From the main HSI page use the rocker pad up/down to increase/decrease the speed. The speed will be displayed in the **Speed** field. Select 100 knots.

Simulator: Be careful when you use the rocker pad to set the speed that you only press the rocker pad up/down. If you inadvertently press it left/right the GPS will think you want to manually steer the simulator and take the GPS out of Auto Track mode. If this happens you will find the GPS won't track towards Cranfield and you will need to stop and then restart the Simulator as described earlier in the chapter.

The simulated plane should now turn onto track and proceed towards Cranfield (EGTC). You should be able to see this most clearly on the Map page - a line will appear from the point where you pressed GOTO to EGTC. You can use the PAGE button to page through the different screens in the main page loop and see how they are updating in real time as you watch.

Airspace

You should immediately start to see some messages appear. On the example route you will soon be alerted with messages about approaching and being inside airspace. Not only is Denham in the Heathrow zone and under the London TMA, but the route takes you straight through Luton's zone.

This teaches a very important point about the Goto function - it takes you straight to your waypoint from your current position no matter what airspace or hazards are in between.

 **Alert:** Never use the Goto function for real without knowing exactly where you are and referring to a current aeronautical map. As you will see later you cannot rely on the inbuilt Jeppesen database for airspace notification in all circumstances.

Have a look at what airspace you are near. Activate the NRST function:

Nearest Airspace			
GPS 92	**GPS III Pilot**	**GPSMAP 195**	**GPSMAP 295**
Acknowledge the airspace messages by pressing [Page] then press [Goto/NRST] <u>twice</u> to activate the **NRST** function. The Airspace nearby will be displayed.	Acknowledge the airspace messages by pressing [Enter] then press <u>and hold</u> [Goto/NRST] to activate the **NRST** function. The Airspace nearby will be displayed.	Acknowledge the airspace messages by pressing [Page] then press [NRST] to activate the **NRST** function. The Airspace nearby will be displayed.	Press [NRST] in response to the prompt. The Airspace nearby will be displayed.

Press [QUIT] to return to the main pages.

Airspace warnings

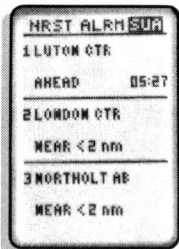

GPS 92

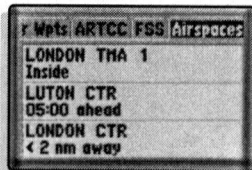

GPS III Pilot

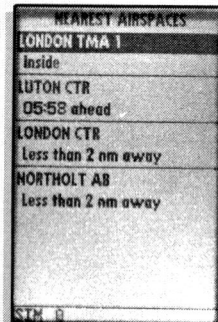

GPSMAP 195

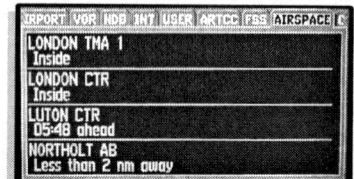

GPSMAP 295

Tuning the airspace warnings

You can tune the airspace warnings. There are two schools of thought about airspace warnings given that UK airspace is far more congested than the USA. One is to turn off all the warnings and rely on conventional navigation (e.g. a conventional, current aeronautical chart). The second is to use the airspace warnings in conjunction with conventional navigation.

Tip: The Author advocates the second approach, because although the Jeppesen airspace database has weaknesses, it acts as a backup and can occasionally spot something you might have missed. It can be especially useful if you are flying in France or elsewhere in Europe in spotting and advising of military airspace and their controlling frequencies.

Alert: Don't try to use the NRST function on the GPS as the sole method to make sure you don't infringe airspace. At the time of writing quite a lot of important airspace in the UK is missing from the Jeppesen database and there is speculation in many quarters that it is exactly this kind of reliance on GPS that is leading to an increase in infringements of controlled airspace by private aircraft.

Airspace Alarms

GPS 92	GPS III Pilot	GPSMAP 195	GPSMAP 295
From the Main Menu, select **Alarms.** You can now edit the settings.	From the Main Menu, select **Setup** and then the **Airspace** tab. You can now edit the settings.	From the Main Menu select **Setup Menu** and then **Alarms.** You can now edit the settings.	From the Main Menu, select the **Airspace** tab. You can now edit the settings.

You can now enable or disable alarms for each type of airspace. The Class B and Class C types also include CTA and TMA respectively. MOA stands for Military Operations Area and SUA stands for Special Use Airspace (in the UK these include Danger areas and Areas of Intense Aerial Activity). Mode C Veils refers to airspace where a mode C transponder is required.

Airspace Alarms

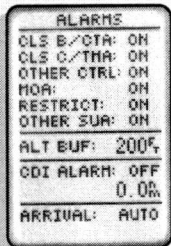

GPS 92

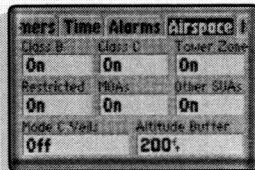

GPS III Pilot

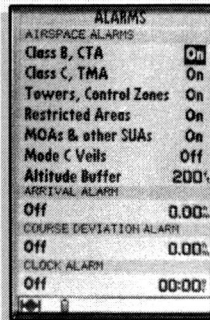

GPSMAP 195

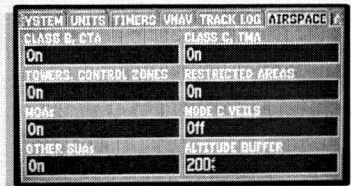

GPSMAP 295

This book assumes all the alarms remain in their default setting of On, except Mode C Veils. This is a matter of personal preference as long as you recognise the consequences of disabling them. Changing the Alarm settings only affects the alerting messages - it does not affect whether the airspace appears on the map.

The alarms also use the Altitude Buffer setting. The Altitude Buffer provides a margin of error or buffer zone around the vertical limits of controlled airspace. This would be good if it wasn't for the inaccuracy of the GPS altitude readout you learned about in Section 1. Since selective availability was turned off in May 2000, a reasonable setting for this is 100 ft. If you set it much higher you may get a lot of spurious alerts.

Vertical Navigation

One message you will get as you approach your destination is "Approaching VNav Profile". Notwithstanding the issue with Altitude inaccuracy the GPS can be configured to alert you when you should be starting your descent towards the destination aerodrome. Ignore this message for the time being. Chapter 13 includes an explanation of how to use the Vertical Navigation Profile.

Choosing your Next Waypoint

By now you may be getting to Cranfield, or even have gone past it. If you still have a little way to go, you can speed up the Simulator to get there sooner.

Once you reach the destination the Simulator continues in a straight line.

Once you have passed Cranfield, enter a new waypoint. Use Oxford for the next part of the exercise.

Quite often you may not know the four letter ICAO designator for an airfield. The GPS knows the full names of all the waypoints in its database. You can use this feature to look up airfields without needing to revert to other references:

GPS 92

Press the [GOTO NRST] button to get to the Goto page. Now press the [WPT] button to get to a more comprehensive waypoint selection page. Highlight the waypoint type at the top left and press [ENTER]. Use the rocker pad up/down to select APT (airport) and press [ENTER] to confirm. Make sure the Location information is shown (if necessary select Location at the bottom of the screen and press [ENTER]. Next use the rocker pad to select the text field below the ICAO field. This is the "Facility" field. Start to enter Oxford letter by letter. Unfortunately the first match clearly isn't in the UK*. Instead, try the next field down, which is the "City" field. Press the [QUIT] key to exit the

Facility field and then select the City field and enter Oxford there instead. This time you quickly discover that the facility name is Kidlington and the ICAO code is EGTK. Press ⬤ENTER to accept it. Now make sure that Done is highlighted on screen and press ⬤ENTER again. You will be taken back to the standard Goto screen and can press ⬤ENTER again to confirm.

GPS III Pilot

Press the ⬤ button to get to the Goto page, and use the rocker pad to make sure the Spell 'n Find tab is selected in the sub menu. Next select the Facility field and start to enter Oxford letter by letter. You only need to get as far as the X when Oxford appears. Unfortunately it's the Oxford NDB*, whereas you were after the airfield. Try the City field instead. Press the ⬤ key to exit the Facility field and then select the City field and enter Oxford there instead. This time you quickly discover that the facility name is Kidlington and the ICAO code is EGTK. Press ⬤ to accept it.

GPSMAP 195

Press the ⬤GOTO button to get to the Goto/waypoints page, and use the rocker pad to select the text field below the ICAO field. This is the "Facility" field. Start to enter Oxford letter by letter. Unfortunately the first match clearly isn't in the UK*. Instead, try the next field down, which is the "City" field. Press the ⬤QUIT key to exit the Facility field and then select the City field and enter Oxford there instead. This time you quickly discover that the facility name is Kidlington and the ICAO code is EGTK. Press ENTER to accept it. Now make sure that Goto is highlighted on screen and press ENTER again.

GPSMAP 295

Press the ⬤NRST button to get to the Goto/waypoints page, and use the rocker pad to select the text field below the ICAO field. This is the "Facility" field. Start to enter Oxford letter by letter. Unfortunately the first match clearly isn't in the UK*. Instead, try the next field down, which is the "City" field. Press the ⬤QUIT key to exit the Facility field and then select the City field and enter Oxford there instead. This time you quickly discover that the facility name is Kidlington and the ICAO code is EGTK. Press ENTER to accept it. Now make sure that Goto is highlighted on screen and press ENTER again.

*The exact results may vary depending on the precise issue of the database.

Using the City & Facility Fields

GPS 92	GPS III Pilot	GPSMAP 195	GPSMAP 295

Alert: Choosing Oxford as the new destination reinforces still further the dangers of relying on GPS alone. Look on a current 1:500000 chart. Goto Oxford direct from Cranfield will take you through R214, Bicester (intense gliding) and Weston on the Green (D129, Parachuting). Bicester also isn't in the Jeppesen database.

Track Log

You may see a dotted line (solid on the GPS 92) following the plane on the map screen. This is your track log and shows where you have flown. By default it is switched on. You will learn more about the Track Log in Chapter 17.

Making the Map More Readable

European countries are much smaller than the United States and the map can look very cluttered at times when the default map settings are in operation. Conversely many of our airfields are considered too small to be displayed except at quite high levels of zoom! There are several ways that you can try to make it more suitable. Whilst your simulated plane is on its way to Oxford (or wherever you chose to send it) you can take a look at some of them.

You can edit the way in which almost every feature of the map will be displayed, so you can fine-tune the display to make it as readable as possible.

Changing the Map Settings			
GPS 92	**GPS III Pilot**	**GPSMAP 195**	**GPSMAP 295**
The GPS 92 doesn't have a base map (roads, railways, rivers and towns) but it does have aviation data, though the setting changes you can make are more limited than on the other models. If you select the **OPT**ions field on the main Map page you can edit the general map setup and also the zoom settings of waypoint and airspace data.	From the Map page press [Menu] to bring up the context sensitive menu and select **Setup Map.** You can now see a number of options spread over several sub-menu tabs.	From the Map page press [Menu] to bring up the context sensitive menu. You can now select from three sub menus to change general **Map Features, Aviation Data** and the base map **Land Data.**	From the Map page press [Menu] to bring up the context sensitive menu and select **Setup Map.** You can now see a number of options spread over several sub-menu tabs.

Map Settings

The Zoom setting indicates the maximum zoom that an item will appear. For example something that has a Zoom setting of 5nm will appear if the map scale is set to 5nm (and below) but won't appear at a setting of 8nm (and above).

You can also set text sizes for most features.

Which settings you choose is a matter of personal preference, though you will almost definitely want to do the following two:

• Increase the visibility of small airports by increasing their zoom setting.

• Increase the visibility of NDBs by increasing their zoom setting.

The Author's personal preferences are shown in Appendix B and you may wish to copy them. Alternatively, here are some pointers:

- Set most text sizes to Small. Avoid using Large text for anything.
- Set Active Route waypoint text to Medium.
- Consider turning Intersections Off.
- Consider thinning roads out by decreasing their zoom setting.
- Consider increasing the visibility of railways by increasing their zoom setting.
- Leave the latitude and longitude grid turned Off.
- Leave Autozoom Off. (Experiment with it - most people find it irritating!)
- On the GPS III Pilot and GPSMAP 195 set the Land Data (base map data) to be Gray rather than On.

Fact: An intersection is a "virtual" navigation point used in Instrument Flying. Very few intersections are of interest to VFR pilots. Perhaps one notable exception is ORTAC at N50° W002°, which defines the point where airway N866 joins the Jersey Zone.

Comparison of settings on GPSMAP 295

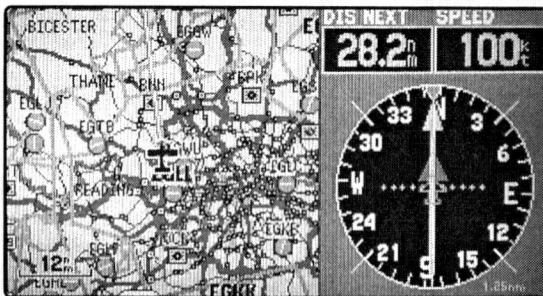

Default settings - very cluttered

Author's preferences

The GPSMAP 295 has a number of preset combinations under the Map Detail setting, however you may still prefer to adjust the settings manually. To have an item adjusted by the Map Detail level set its Max Zoom to AUTO.

"Goto" and Simulator Mode | Comparison of Map Settings

Quick De-clutter Option (GPSMAP 295 only)

If you have the GPSMAP 295 you can quickly remove layers of detail from the map as follows by pressing the **ENTER** button on the map screen:

- Press once to remove all the base map information (roads, railways, rivers and towns etc). Clear-1 appears by the map scale indicator.
- Press again to remove all the airspace markings. Clear-2 appears by the map scale indicator.
- Press **ENTER** again to remove everything except the waypoints and legs in the active route. Clear-3 appears by the map scale indicator.
- Press **ENTER** again to display everything.

Switching the Simulator Off

There are two ways of switching the Simulator off.

The easy way is to turn the GPS off. As a safety feature the Simulator always switches off when the GPS is switched off, so that you don't forget and use it for real navigation next time you switch on. Otherwise you can turn the Simulator off manually:

Switching the Simulator Off			
GPS 92	**GPS III Pilot**	**GPSMAP 195**	**GPSMAP 295**
From the main menu select **Setup Menu** and then **Operation.** Set the **Current Mode** to **Normal.**	From the Satellite Status screen press [Menu] once and select **Stop Simulator.**	From the Satellite Status screen press [Menu] once and select **Exit Simulator.**	From the Satellite Status screen press [Menu] once and select **Stop Simulator.**

Summary

This has been a pretty intensive Chapter, but in it you've finished learning all the basics about using your GPS. You are now ready to go flying. The next Chapter looks at installing the GPS in the aircraft and then in Section 3 you can apply what you've learned to real world navigation.

Introduction

In this Chapter you will learn how to install the GPS in the aircraft.

Before you go flying you will want to have completed all the basics in this Section of the book. If you are going to fly a route for real you will also want to go through Section 3 on Flight Planning too.

> **Airmanship:** If you are going to try out your GPS in the air for the first time, always take a safety pilot to help with lookout (or to actually fly the plane) as you will probably spend far longer with your head in the cockpit than usual.

The Garmin range of Portable GPS receivers can all be yoke mounted. The GPS III Pilot and the GPSMAP 295 can also be "dash" mounted on the glare shield (or in a car!).

Where you have the choice, whether to yoke mount or dash mount is a matter of personal preference. Many people dislike dash mounting for four reasons: firstly, dash mounting obscures too much of the forward view over the nose of the aircraft and secondly it can be nearly impossible to operate the GPS in bumpy conditions. Thirdly, you are more likely to get glare on the GPS screen if it is dash mounted and find it more difficult to shield that glare when it occurs. Finally, the dash mounts require sticking to the glare shield, which is not normally an option in hire aircraft.

GPS III Pilot mounted on yoke and glare shield. Note the glare on the lower picture even on an overcast, rainy day.

Of course if you have a joystick rather than a yoke then you will have to mount the GPS on the glare shield. If you own the plane then you can even consider other mounting options - the author has seen a Europa installation with a GPSMAP 295 mounted flush in the dashboard.

The yoke mount is an optional extra with the cheaper units, but included with the GPSMAP 195 and GPSMAP 295.

The rest of this Chapter assumes you are going to yoke mount the GPS.

Yoke Mounting the GPS

The first thing to consider is which side to mount the GPS. This isn't as daft as it sounds. If you are operating as a solo pilot or with "non-flying" passengers then the choice is made for you (left hand seat!), however if you are flying as P1 and sharing the flying with a fellow PPL who is navigating from the right hand seat, you may well wish to install on the right hand side. This has the advantage that no important instruments are obscured and the risk of interference to your hands on the controls from wayward cables is removed. Of course you need to make sure your navigator can use the GPS!

You should first install the yoke mount and then the antenna.

Installing the Yoke Mount

Install the yoke mount in accordance with the user manual. Most yokes have two screw knobs - one for setting the angle of the cradle that holds the GPS and the other which clamps the mount to the yoke. You may need to "offer up" the mount to the yoke and pre-position the angle of the cradle before you can get it to fit properly. If you fly the same aircraft type all the time, you shouldn't have to change it again.

Some yoke mounts come with two optional knobs for the clamp, one with a long screw thread and one with a short one. If yours is one of these then use the shorter one if possible as it prevents the knob hanging down too far (this can catch on your knees or the cables).

Checklist: Install the yoke and antenna at the end of the "A Check" (or transit check as applicable). That way you aren't needing to fiddle about with it whilst loading and briefing passengers.

Tip: You can't fit the GPSMAP 295 on the new style yoke of recent Piper aircraft as it is too wide to fit between the "horns" (luckily, most such aircraft the author has seen come with the Garmin GNS 430 panel mount GPS NAV/COM!).

Installing the Antenna

Next install the antenna. You have the choice of mounting the antenna on the GPS receiver itself or remote mounting the antenna using either a remote mount kit or a complete remote low profile antenna. (These items are optional extras on the cheaper units, but standard with the GPSMAP 195 and included on the GPSMAP 295). The GPSMAP 195 actually has an antenna built in so may not need the external extension.

> **Alert:** If you have yoke mounted the GPS it is recommended that you use the remote antenna as it is far less likely to get obscured from the satellites than a locally mounted antenna.

Position the remote antenna on the windscreen and stick it on using the suction cup(s) on the mount.

If this unsticks it will fall on the floor!

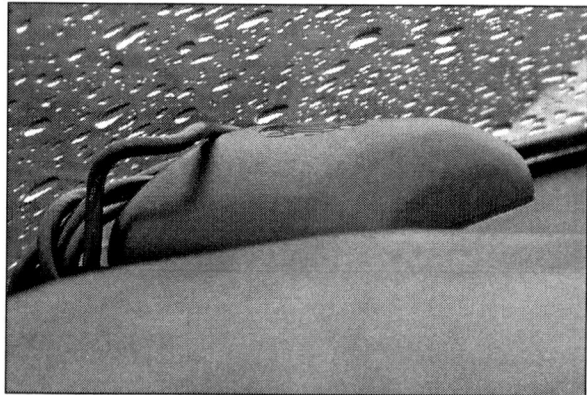

Nowhere to fall!

> **Tip:** You don't have to stick the remote antenna high up on the windscreen - if you stick it where the windscreen meets the glare shield then even if it un-sticks in flight, it has nowhere to fall.

You can put a couple of coils of cable around the yoke to make sure that if the cable is pulled in flight the connection to the GPS receiver is not damaged. You should also ensure that any unused cable is coiled up and secured so that it cannot foul the controls in flight.

Installing the Power Cable (if required)

Run the external power cable from the cigarette socket (if fitted) to the yoke making sure that the plug is secure in the cigarette socket to minimise the risk of it coming

GPSMAP 295 installed in a PA28.

loose in flight. Like the antenna cable, you can put a couple of coils of cable around the yoke to make sure that if the cable is pulled is flight the connection to the GPS receiver is not damaged. You should also ensure that any unused cable is coiled up and secured so that it cannot foul the controls in flight.

Magnetism: Don't run the power cable near the compass. Electric wires create a magnetic field when current runs through them, which will affect the compass.

Alert: Unfortunately, many aircraft have the cigarette socket on the right hand side of the aircraft. You should seriously consider operating the unit on batteries if you cannot route this cable safely.

Tip: Note that on some Garmin yoke mounts the power cable is attached through a cable clamp. If you are not going to use the power cable you should detach it in accordance with the manual.

Installing the GPS

First double-check the GPS has batteries (even if you intend to run it off external power). You should also check that the battery power meter is configured for the type of batteries you are using (as discussed in Chapter 8):

Getting to the Aircraft **Installing the GPS**

Now clip the GPS into the yoke mount following the instructions in the manual. Depending on the yoke mount this may automatically engage the power connection, otherwise plug in the power cable (if required).

Referring to the manual, make sure the GPS is securely clipped in and secured depending on the model (for example the GPSMAP 195 has a twist knob below the yoke mount and the GPSMAP 295 has a little screw knob behind top centre).

Finally, plug in the antenna.

"Full and free" check.

Checking for Full and Free Movement

You need to make sure that the cables and the yoke don't obstruct full and free control movement. Gently move the controls now checking for full and free movement. You want the cables to allow for full movement to the yoke, whilst not being so slack that they get in the way when the yoke is at rest. Also check that the clamping bolt doesn't catch on your knees or kneeboard.

Checklist: Remember to check this at all the points where you do a full and free check in your current checklist.

Contrast and Brightness

You may wish to change the contrast or brightness/backlighting of the screen as described in Chapter 8. Though the default settings are normally OK for most purposes, now the GPS receiver is in-situ you might need to make an adjustment.

Tip: If you find you need to adjust the GPS 92, GPS III Pilot or GPSMAP 195 contrast in flight, use up/down on the rocker pad with the Satellite Status page displayed to activate the adjustment control, and left/right to make the adjustment - then press ⓔⓝⓣⓔⓡ.

Alert: Selecting the backlight to be always on can significantly reduce battery life.

VFR Flight Planning using GPS

In this section you are going to add what you have learned so far, step by step into the flight planning and in-flight process. In addition you will progress your knowledge beyond the use of the simple Goto button and into planning full routes. You will also find out about some of the more advanced options to take the chores out of flight planning.

The section makes extensive use of the Simulator to help you work through the examples on the ground before going airborne for real.

You will need access to all the usual tools of VFR flight planning including current aeronautical chart, ruler, protractor, writing implements, a VFR flight log, current airfield reference guide (such as Pooley's or Jeppesen), Pooley's CRP-1 flight computer or equivalent ("wizzwheel"), plus of course your trusty GPS with adequate supplies of batteries.

Tip: Unfortunately the Garmin GPS receivers do not replace many wizzwheel functions so once you have gained your PPL you might want to buy an electronic "wizzwheel" which looks a bit like a pocket calculator and can do the wind calculation in about half the time of an experienced wizzwheel user. Carry your original wizzwheel with you in your flight case and remember how to use it though, as sooner or later your electronic equivalent will run out of batteries!

This section assumes that you are fully conversant with finding your way around all the GPS functions covered so far especially looping through the main pages and accessing the Main Menu, as well as using the in-built Simulator.

Introduction

This Chapter does a recap of conventional VFR flight planning and uses Goto on a Simulated flight.

It is assumed that you are already well versed with the basics of VFR flight planning. If you aren't then there are many good texts on the subject including Book 3 in Trevor Thom's Air Pilot's Manual series and Book 3 in the AFE Private Pilot's Licence Course written by Jeremy Pratt.

> **Alert:** This book superimposes the benefits of GPS onto VFR flight planning and navigation using Dead (or "deduced") Reckoning (DR). You should not throw conventional planning away! If you do, then you will probably get lost as soon as GPS fails (and it will!).

There may well be differences in your approach or the way you have been taught to do flight planning. This book has adopted the commonly accepted principles. You should continue to use the methods you have been taught rather than try and change to exactly match the Author's approach, and just be aware of the differences as you work through the next few Chapters.

As well as recapping on VFR planning, this Chapter also teaches you more about using the Goto feature you first learned about in Chapter 9 in a straightforward VFR flight.

Many newcomers to GPS use the Goto feature on the basis that it is a lot easier to use than putting in a full route plan. This is only true if you are using an unfamiliar GPS. Once you become familiar with your GPS you will discover that doing a full route plan is actually very much easier than using Goto, both on the ground and in the air - in fact as you work through the next few Chapters you will discover that Goto can be really quite cumbersome by comparison.

You can do this Chapter as "ground school" and fly it using your GPS receiver's Simulator and then use what you've learned to go on to do a flight in real life.

Your first trip will be a fairly short flight from Elstree (EGTR) to Kemble (EGBP). Use a 1:500000 scale Southern England and Wales chart.

Although this is a short flight you will be planning it well as if for real. You will also not be using such a direct a route as you might normally so you have time to try out the GPS on a few waypoints.

For this flight you'll be using just what you've learned so far. You won't be using the full flight planning capabilities of the GPS until the next Chapter.

1. Choose the Waypoints.

Choose each leg and draw it on an aeronautical chart. As you choose each leg you will be looking for obvious pitfalls such as areas it may be impossible or difficult to get a clearance through such as restricted or prohibited areas, Class A airspace, or (increasingly) class D around major international airports. Also look to avoid high ground if the weather is likely to be poor.

Choose a route between two easily identifiable waypoints. This is the first time you may encounter a slight dilemma - the GPS has airfields and navaids (such as VORs and NDBs) in its database. Using VFR navigation you may often wish to use other landmarks such as towns or distinctive bridges etc. Luckily many navaids are on active or disused airfields, which are still valid VFR landmarks. Later on you will learn how to set up user-defined waypoints so that you can define other waypoints you regularly use.

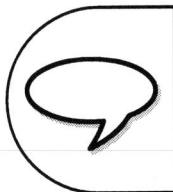

> **Tip:** This is where the GPSMAP 295 has a huge advantage over the less expensive models - it has a database of thousands of towns and cities built in. It even has some motorway intersections. If you add a MapSource cartridge and CD you can even specify a street address as a waypoint!

For our example route we will go from Elstree to Bovingdon, then from Bovingdon to Compton VOR (Hampstead Norris) and then direct to Kemble.

Bovingdon has the advantage that it is a navaid on a very identifiable disused airfield, and Compton that it is on a small grass airfield. This means they make fairly good visual waypoints as well as being in the GPS database.

Now write the waypoints on your flight log. If you are using a student's flight log you may wish to write in the waypoint type and "observation", i.e. what you expect to see when you reach the waypoint.

FROM / TO	MSA	PL/ALT	TAS	TR (T)	W/V	HDG (T)	HDG (M)	G/S	DIST	TIME	ETA	ATA
ELSTREE / BOVINGDON					/							
BOVINGDON / COMPTON					/							
COMPTON / KEMBLE					/							

2. Minimum Safe Altitude

Now compute the minimum safe altitude. There are several ways this is taught. Many people favour looking for the highest object 5 miles either side of track, rounding up to the next 100 feet and adding 1000 feet. You may have been taught something different (such as looking ten miles either side of track or adding 1500 feet to the maximum elevation figures on the chart) in which case stick to what you know best. Enter the MSA on your flight log.

The highest MSA on the example route looks to be about 2200 feet (using the 1000 feet method).

3. Plan your Altitude

Select the altitude you will fly at taking into account the MSA, airspace and so on. Enter this on your flight log.

For the example route fly initially at say 2300 feet. This will keep you below the London TMA. Once clear you can climb higher if you wish.

4. Measure the Leg Distance

Now measure the distance of the leg, making sure of course that the ruler scale matches the map scale. Enter the distance on your flight log. In Chapter 13 you will learn how to measure the leg on the GPS and use the ruler as a gross error check.

5. Measure the True Track

Measure the True Track between the waypoints using the protractor. Enter this on the flight log. When you get onto using the GPS to help with full flight planning in Chapter 13 you will find out how to get the GPS to do this measurement.

6. Compute your True Airspeed

Computing your true airspeed for the pressure altitude you intend to fly at is normally done using a "wizzwheel" using the information found in Met form 214.

In this example you will be flying a PA28 with an indicated airspeed (IAS) in the cruise of 100 kts. Assume that the Met form 214 for the day is showing the temperature is 10 degrees at 2000 feet and 4 degrees at 5000 feet. By interpolation it is going to be 9 degrees at 2500 feet.

You could use your wizzwheel to do the airspeed computation, however the GPS has a feature that will do the calculation for you - you need to use the Density Altitude function that will compute not only the density altitude but also the True Airspeed (TAS).

Tip: To be honest, the fact that the GPS can compute the TAS is only included here for completeness - it probably takes less time to compute it using a wizzwheel.

Computing True Airspeed (TAS)			
GPS 92	**GPS III Pilot**	**GPSMAP 195**	**GPSMAP 295**
From the Main Menu select the **E6-B Menu.** Enter in 2500 feet as the Indicated Altitude **(IAlt)** and 100kts as the Calibrated airspeed **(CAS).** Enter 9°C as the **Total Air Temp.** If you know it, enter the QNH in the pressure setting **(Prs)**, otherwise use 1013. The TAS is 103kts.	From the Main Menu select **E6B** and **Density Alt.** Enter in 2500 feet as the **Indicated Altitude** and 100kts as the **Calibrated airspeed.** Enter 9°C as the **Total Air Temp.** If you know it, enter the QNH in the **Pressure** setting, otherwise use 1013. The TAS is 103kts.	From the Main Menu select **Density Alt/Winds Aloft.** Enter in 2500 feet as the **Indicated Altitude** and 100kts as the **Calibrated airspeed.** Enter 9°C as the **Total Air Temperature.** If you know it, enter the QNH in the **Pressure** setting, otherwise use 1013. The TAS is 103kts.	From the Main Menu select E6B. Enter in 2500 feet as the **Indicated Altitude** and 100kts as the **Calibrated airspeed.** Enter 9°C as the **Total Air Temperature.** If you know it, enter the QNH in the **Pressure** setting, otherwise use 1013. Ignore the wind related information. The TAS is 103kts.

Fact: If you wanted to be really accurate you could have consulted the aircraft's POH to work out its calibrated airspeed (CAS) from its indicated airspeed (IAS) and then used the CAS to find the rectified airspeed (RAS). You would then use the RAS in the computation instead of the IAS. At the speeds involved here the differences are very small.

7. Compute your True Heading

Using your wizzwheel, compute your True Heading, i.e. the heading you will need to steer to achieve the True Track. Unfortunately this is one really useful function that the Garmin GPS receivers lack, even though, as you will see later, they can compute the actual wind aloft during a flight! You will need the other information on Met form 214 - the forecast wind aloft.

For this exercise use 15 knots from 270 degrees (270/15).

8. Compute your Magnetic Heading

Adjust the True Heading for Magnetic Heading by adding (or subtracting) magnetic variation. Magnetic variation varies over time and at the time of writing in the UK you are likely to have to add between 3 and 6 degrees of variation ("Variation West, Magnetic Best").

In the example the magnetic variation is around 3.5°W, so add, say, 4 degrees.

For completeness you should also take account of any compass deviation as shown on the compass deviation card in the actual aircraft you are going to fly.

9. Compute your Time

Using your wizzwheel again divide distance by speed to get the time. Round the time to the nearest minute and enter it in your flight log. In Chapter 13 you will learn how to use the GPS to do this computation.

FROM / TO	MSA	PL/ALT	TAS	TR (T)	W/V	HDG (T)	HDG (M)	G/S	DIST	TIME	ETA	ATA
ELSTREE / BOVINGDON	1800	2300	103	297	270/15	293	296	89	9	6		
BOVINGDON / COMPTON	2200	2300	103	241	270/15	245	248	90	29	19		
COMPTON / KEMBLE	2200	2500	103	290	270/15	287	290	89	33	22		

10. Compute your fuel requirement

Compute the fuel requirement for each leg based on the stage of flight and the published fuel settings for your aircraft. In Chapter 13 you will learn how to use the GPS to do this computation. Remember to add Taxi fuel (normally a couple of gallons/10 litres) and holding/diversion time (typically 30-45 minutes for a VFR flight) to get the full requirement.

ASCO				
VFR FLIGHT LOG				

DATE / /		FUEL CONSUMPTION	*10/h*	50
PILOT	AIRCRAFT	TOTAL REQUIRED		
FROM	TO	FUEL ON BOARD	*8*	
DISTANCE	FLIGHT TIME	RESERVE	*50*	
START UP	TAKE OFF	TOTAL ENDURANCE	*10*	
LANDING	SHUT DOWN	COMMUNICATIONS & RADIO NAV. INFORMATION		
ALTERNATE		DISTRESS 121.50		40
DISTANCE	FLIGHT TIME	STATION	SERVICE	FREQUENCY

11. Fan Lines ("track guides")

If you wish, you can now add fan lines to the chart. Fan lines are drawn at typically 5° or 10° to the track and can be used to help you estimate your track error if you go off course. A lot of people leave off fan lines once they have obtained their PPL and become more experienced. This decision is up to you. Later on you will find out at how the GPS can help you get back on track.

Ten degree lines have been added to leg 1 in the example.

12. Track Markings

You can now add distance or time marks. There are typically three different methods taught.

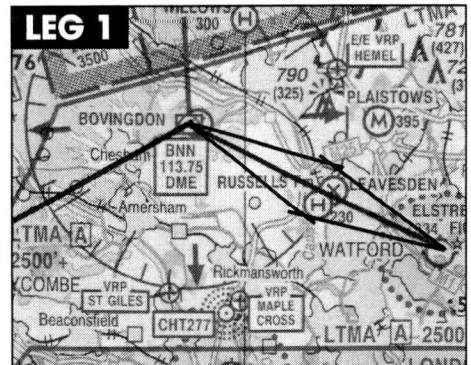

● **6 minute markers.** These are drawn every six minutes down the track and are easier to draw than they sound since 6 minutes is a tenth of an hour, and therefore you simply divide the speed for the leg by ten to get the distance travelled in 6 minutes and draw a marker at each increment. In flight planning this provides a gross error check on both the time and leg measurement and in the air enables time since the last waypoint to be used very easily to give a rough position.

10 mile markers. These are drawn every 10 nautical miles down the track. They can be used to very quickly identify probable position in conjunction with DME or GPS distance information. They also make it easier to use the one in 60 rule.

Fractional markers. These are drawn to segment the leg into quarters and help apply a useful rule of thumb based on the one in 60 rule to correct track error.

> **Tip:** The Author is in favour of 6 minute markers as they add time information visually into the navigation picture which the other methods don't (it takes a lot more mental agility to project time onto a map when airborne than either distance or fractions). In addition they provide gross error checking of the pre-flight measurements and computations which other methods don't. You should use the one you already know rather than try to change.

You may prefer to add mileage markers or fractional markers as you prefer.

Finishing off...

All you would normally have to do now, having checked the weather, is work out who you will talk to on the radio and enter all the frequencies, get the plates handy for your departure and destination airfields and you are planned and ready. You might also want to consider an alternate.

In the example, you are ready with the frequencies for Elstree, Wycombe Tower, Benson Radar, Brize Radar, and Kemble. You've got the airfield plates for Elstree and Kemble and your alternate.

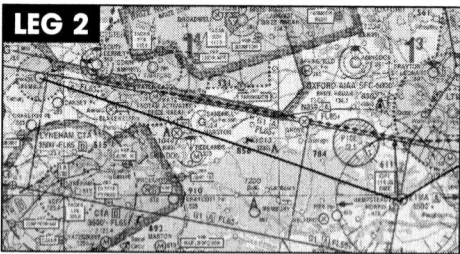

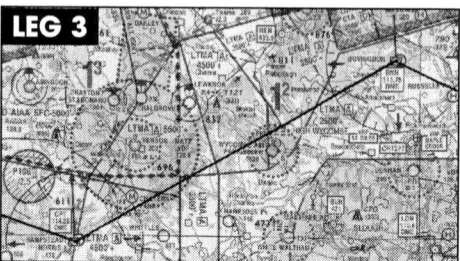

Flying the Route on the Simulator

As this is your first planned flight, fly it on the Simulator before you go airborne.

Set the GPS position so the Simulator starts in the right place - you can start the Simulator as described earlier or use the slightly quicker method described below:

Freshly power up the receiver. It doesn't matter if you are inside or whether there is an antenna connected:

Start the Simulator			
GPS 92	**GPS III Pilot**	**GPSMAP 195**	**GPSMAP 295**
From the Main Menu select **Setup Menu** and then **Operation**. Set the **Current Mode** to **Simulator** and the **Initial Position** to EGTR (with a **Bearing** of 0 and **Distance** of 0).	From the Satellite Status screen press [Menu] once and select **Start Simulator.** Next select **Initialize Position.** Zoom in and highlight Elstree (EGTR) north east of Heathrow (EGLL) and press [Enter].	From the Satellite Status screen press [Menu] once and select **Start Simulator.** Next select **Initialize Position.** Zoom in and highlight Elstree (EGTR) north east of Heathrow (EGLL) and press [Enter].	From the Satellite Status screen press [Menu] once and select **Start Simulator.** Next select **Initialize Position.** Zoom in and highlight Elstree (EGTR) north east of Heathrow (EGLL) and press [Enter].

Note: Elstree is classified as a small airport. If you didn't adjust the default map zoom settings as suggested in Chapter 9 then you may have to zoom in considerably before it appears.

Elstree to Bovingdon

Goto BNN (the VOR on the disused airfield at Bovingdon). You will discover that there is a BNN in the UK and Norway. You can use the rocker pad up/down to select the one you want. Luckily the cursor is already on the UK BNN so you can just press (ENTER) .

Alert: Forgotten how to use Goto? Lucky you aren't in the air yet! You can refresh your memory by referring to Chapter 9.

Imagine that you are taking off.

🌙 On the HSI page use the rocker pad up/down to accelerate the simulator to 90 knots to approximate your forecast speed (the GPS only lets you set the speed in 10 knot increments). GPS 92 users should edit the speed on the CDI page.

🌙 Use (PAGE) to reach the Navigation/Position page (which has the clock display) and note your takeoff time on the flight log.

🌙 Set the Simulator Altitude to 2300 feet.

🌙 Go back to the map page and zoom to a suitable scale.

As you exit the imaginary circuit the plane should turn on track towards Bovingdon.

> **Simulator:** It's very important when setting the speed using up/down on the rocker pad not to apply left or right pressure as this will cause the Simulator to enter manual steering mode and it won't automatically follow the track. If you do you will need to restart the Simulator. (On the GPS III Pilot and GPSMAP 195 you can reset the Simulator into Auto Track mode).

> **Getting left behind?** Remember that you can slow down the speed of your virtual plane at any time from the HSI screen (CDI screen on GPS 92) - just remember to take this into account on your time computations.

Compute your ETA at Bovingdon using the flight log.

As you approach Bovingdon you can think about getting ready for the next leg. You should do all the things you normally do and add in the GPS - many people use the "HAT" mnemonic before and after the turning point.

Before the turning point:

H : Think about the Heading for the next leg.

A : Think about Altitude and the Airspeed.

T : Prepare to mark the Time of arrival (ATA) on the flight log.

After the turning point:

H : Check the Heading (and DI/Compass alignment).

A : Check the Altitude and Airspeed.

T : Compute the Time (ETA) at the next waypoint.

You can set up the next Goto as you do your first HAT check (or equivalent) as you approach the waypoint. Press (GOTO) and enter in CPT but don't press the (ENTER) key to activate it yet - wait until you get overhead your first waypoint. In real life you will be able to see the waypoint and get the timing just right. In Simulator mode you can't see the map and the Goto page at the same time so you will need to just get it as close as you can for the purposes of this exercise.

> **Alert:** It is important to realise that Goto takes you from where you are to the waypoint you specify, not from where you think you are or where you'd like to be! When you use Goto, if you aren't on your planned track then your whole flight log becomes inaccurate and you risk clipping airspace you planned to avoid, compromising your MSA and just plain getting in a muddle as the GPS track no longer matches your planned track. That's why it's important to activate Goto overhead the waypoint.

Bovingdon to Compton

For simulation purposes you can leave the speed at 90 knots to approximate your forecast speed.

As well as the HAT checks you might also do "Cruise Checks" before or after a waypoint. Many people use the FREDA mnemonic:

- **F**uel
- **R**adio
- **E**ngine
- **D**I
- **A**ltimeter

> **Airmanship:** Aviate, Navigate, Communicate. Then and only then fiddle with the GPS! Once it becomes second nature it will naturally fit in with Navigate - remember to look out of the (imaginary) cockpit!

Airspace Warnings

By now you will probably have encountered some Airspace warnings. You can take a look and see the closest airspace - activate the NRST function. You will probably see Booker, Oxford and Benson.

If you select Booker (Wycombe) you can get more information. You can select **Frequencies** to find out the controlling frequencies - you can see that the tower frequency is 126.55 (at the time of writing).

The other thing you might notice is that the circle on the map looks much bigger than an ATZ. In fact it's a 5nm circle. This is a quirk of the Jeppesen Database.

Jeppesen Database: At the time of writing the Jeppesen database doesn't contain standard UK ATZ information - it puts a 5nm circle around any airfield with an ATC tower frequency.

Next you will see that you are just skimming through the edge of a 5nm circle around Benson. This isn't Benson's MATZ - its just a 5nm circle around Benson's tower - at the appropriate level of zoom you will see that it doesn't have a "pan handle" at the northern side.

Jeppesen Database: The Jeppesen database in Garmin portable receivers doesn't contain any UK MATZ information. (At the time of writing Jeppesen have just introduced MATZ information into some of their other publications so they may be present in the Garmin database by the time you read this).

By now you should be nearly at Compton VOR. Being on a grass airfield, it can be difficult to see, but if you have it in sight you can do your HAT checks and prepare the next Goto to Kemble (EGBP). Activate it when you are in the overhead, not forgetting to aviate, navigate and communicate first. Fiddle with the GPS last!

Simulator: Again, in real life you will be able to see the waypoint and get the timing just right. In Simulator mode you can't see the map and the Goto page at once so you will need to just get it as close as you can for the purposes of this exercise.

Have you noted the ATA - did it work out OK? Have you done your cruise checks?

Compton to Kemble

For simulation purposes, you can leave the speed set to 90 knots. Compute the ETA at Kemble using the flight log information. Compare it to the ETA on the Active Route (Active Goto) page. You are now flying under Airway G1. Again there is no sign of this on the GPS.

> **Jeppesen Database:** The Jeppesen database contains CTA information but not airway information.

Now you are on the last leg, there is time to see another feature of the map page.

If you have the GPS III Pilot, GPSMAP 195 or GPSMAP 295 use the rocker pad to move up and down, left and right. You will see that a pointer cursor appears (a crosshair appears on older GPSMAP 195 receivers). You can point at items and press **ENTER** to find out more about them. Try highlighting the circle just to the west beyond Kemble. It's R105 (Highgrove House), and you can find more about it by pressing **ENTER**. The cursor position is also displayed on screen as well as its bearing and distance from the GPS position. You can pan the screen by moving to its edges.

The GPS 92 isn't quite as flexible as the other models. The rocker pad enables you to select between different database points on the screen. Press **ENTER** for the details. You can pan the screen by selecting the "Pan" field at the top of the page.

To return to normal map operation press **QUIT** .

As you approach Kemble your HSI page (CDI page on the GPS 92) tells you both the distance to go and the time to go (estimated time en-route - ETE). With about 10 miles to go to Kemble you probably need to be calling them on the radio.

By now, you may be getting an "Approaching VNav Profile" message. Ignore this for the time being - you didn't set it up for this flight so you shouldn't be relying on it - use traditional methods of descending into the (imaginary) circuit this time! Chapter 13 introduces this feature properly.

With the airfield in sight you can now concentrate on positioning yourself for landing. You don't need to even think about the GPS now (and anyway the Simulator can't do circuits and landings and will continue on in a straight line until you turn it off!).

Summary

In this Chapter you have learned how to use the GPS to help plan a VFR flight and use the Goto function to help navigate it, as well as how to find airfield frequencies and seek out information about other features on the map.

You may have found it pretty hectic! Ideally you don't want to be entering new waypoints into the GPS each time you reach a turning point, and this means entering the route in full before flight. This is what you will learn next, and you will hardly ever need to use Goto again!

Kemble Airfield

If the last Chapter hasn't been enough to put you off, you now have enough knowledge to fly a route for real using the GPS to aid you!

Like many things in flying, learning something new can tip the balance between being in control and being overloaded, especially if you are recently qualified!

This Chapter is deliberately short with the key points highlighted - otherwise fly the route as you would "normally" using your VFR flight log.

Alert: The first time you use the GPS for real go up with a safety pilot, or better still sit in the right hand seat and do the navigating before progressing to the left hand seat.

Checklist: It's a good idea to switch on your GPS at the same time you switch on your other avionics. This is normally shortly after you have started the engine. Remember to check the database validity and allow sufficient time for the receiver to acquire satellites.

Fly the Heading not the Line: Fly the heading on your flight log, not the line on the GPS. If the wind forecast is correct and you fly accurately this will take you down your desired track. How to correct for track errors is explained in Chapter 16.

Checklist: When you do your cruise checks is a good time to check the GPS is functioning properly. Page to the Satellite Status screen. Do you have good coverage? Is the Battery OK (or are you still on external power)? Does the waypoint the GPS is aiming at match the one you are really aiming at?

Tip: When you are airborne things are very different from on the ground and you may find you temporarily forget some of your GPS training! The following tip works anywhere to get you back to the Satellite Status page: Just keep on pressing `QUIT` . No matter where you are, you will soon be back to the Satellite Status page.

Alert: Once you have the destination airfield in sight, STOP fiddling around with the GPS and make sure you have done your airfield approach checks and in due course your pre-landing checks.

Checklist: As you shut the aircraft down you will probably switch the avionics off. Now is a good time to switch off the GPS.

Introduction

In this Chapter you will use the full power of the GPS by programming in an entire route.

As you discovered in the previous Chapter, Goto is easy to use but is a little cumbersome and open to risks. What you really want to do is maximise the benefit of GPS and have it take as much workload as possible.

In this flight you're going to continue your trip northwards from Kemble up to Caernarfon. Again you will need the Southern England and Wales 1:500000 chart.

Planning the Route

1. Choose the Waypoints. Take a direct route from Kemble to Gloucestershire, Gloucestershire to Shobdon, from Shobdon to Llanbedr, and from there to Caernarfon. Draw the line on the map.

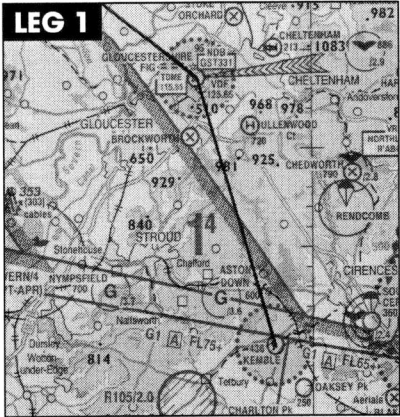

Alert: You are going to have your GPS to stay on track, but even so you might want to be quite generous on the method for computing the MSA as there can be as much as 1100 feet between methods on the last leg!

2. Minimum Safe Altitude

Now check the minimum safe altitudes. There are some very high MSAs as you are passing near some of the highest ground in the UK.

3. Plan your Altitude

Plan to climb to flight level 45. This will give you clearance above the higher ground over the mountains and this is the correct quadrantal level for the direction you are heading. There is no airspace to worry about.

FROM / TO	MSA	PL/ALT	TAS	TR (T)	W/V	HDG(T)	HDG (M)
KEMBLE / GLOUCESTER	2000	CLIMB			/		
GLOUCESTER / SHOBDON	2400	FL 45			/		
SHOBDON / LLANBEDR	4000	FL 45			/		
LLANBEDR / CAERNARFON	4600	DESCENT			/		
/							
/							

Entering the First Route Waypoint

Now you are ready to enter the route into the GPS. The Garmin receivers let you create up to 20 routes of up to 30 waypoints each.

Entering a Route Waypoint			
GPS 92	**GPS III Pilot**	**GPSMAP 195**	**GPSMAP 295**
From the Main Menu, select **Routes**. To choose the route number, check that the route number is highlighted and press [Enter]. You can now use up/down on the rocker pad to choose the route you are going to edit. Select route 1. The route title field is now highlighted. Ignore this for the time being and use the rocker pad to move down and highlight waypoint 1. Now Press [Enter] and use the up/down rocker pad to change the letters and left/right to move along. Enter the ICAO code for Kemble (EGBP) and press [Enter]. The cursor will highlight the next waypoint.	From the Main Menu, select **Routes.** On the Routes page press [Menu] once to activate the context sensitive menu and select **New Route.** The Route Plan page appears and the first waypoint is highlighted. Now Press [Enter] and use the up/down rocker pad to change the letters and left/right to move along. Enter the ICAO code for Kemble (EGBP) and press [Enter]. The cursor will highlight the next waypoint.	From the Main Menu, select **Routes.** On the Routes page press [Menu] once to activate the context sensitive menu and select **Create New Route.** The Route Edit page appears and the first waypoint is highlighted. Now Press [Enter] and use the up/down rocker pad to change the letters and left/right to move along. Enter the ICAO code for Kemble (EGBP) and press [Enter]. The cursor will highlight the next waypoint.	Press the [Route] button. Highlight **New Route** and press [Enter]. The Route Plan page appears and the first waypoint is highlighted. Now Press [Enter] and the waypoints page will appear. Make sure the **Aviation** tab is visible with the top left field highlighted and press [Enter] to start editing. By default you may find the nearest airfield to you is already selected. Use the up/down rocker pad to change the letters and left/right to move along. Enter the ICAO code for Kemble (EGBP) and press [Enter], highlight **Use** on screen and press [Enter] again. The waypoint will be entered into your route. Move the cursor to highlight the next waypoint.

Carrying on with the Route

The chart shows that Gloucestershire has an NDB on it so rather than look up the ICAO code for the airfield itself you can quickly enter in the code for the beacon instead as its code is shown on the chart. Enter GST as waypoint 2.

Once you have entered the waypoint the GPS will automatically display the Desired Track (DTK) (also known as "course"). You can also see the Leg Distance (DST) between the points.

Fact: On the GPS III Pilot you can use the rocker pad to scroll through and display all the data fields. On the GPSMAP 295 you can configure the fields that are displayed.

You might also notice that the route title is being updated as you enter in each waypoint. It now reads EGBP-GST.

Shobdon also has an NDB. Enter SH as waypoint 3. There are quite a few locations with the SH identifier - make sure you select the one in the UK!

Now enter Llanbedr airfield as waypoint 4. It doesn't have a navaid on it but you can get its ICAO code from the panel at the edge of the chart (or from other publications such as a flight guide) - it's EGOD.

Tip: If you've got a GPSMAP 295 you can use the facility field (the third field down) and select LLANBEDR directly instead of looking up the ICAO code.

Finally enter Caernarfon as the last waypoint. You can get its ICAO code from the edge of the chart or, since you will need the plate anyway, from your flight guide - it's EGCK.

Route Pages

| | GPS 92 | GPS III Pilot | GPSMAP 195 | GPSMAP 295 |

GPS 92

ROUTE: 1
EGBP TO EGCK
NO WAYPNT DTK DST
1 EGBP 347° 14
2 GST 314° 34
3 SH 312° 57
4 EGOD 341° 19
5 EGCK
TOTAL DST 124

GPS III Pilot

EGBP-EGCK   New Route
Waypoint ◄ Leg Dist ►
EGBP 14.1
GST 33.7
SH 57.0
EGOD 19.3
EGCK   TOTAL 124

GPSMAP 195

ROUTE EDIT
01 EGBP - EGCK
WAYPOINT DTK DIS
EGBP 347° 14.1
GST 314° 33.7
SH 312° 57.0
EGOD 341° 19.3
EGCK
TTL DIS 124   Done?

GPSMAP 295

ROUTE PLAN
EGBP-EGCK
WAYPOINT
EGBP 0.0
GST 346° 14.1
SH 313° 33.7
EGOD 311° 57.0
EGCK 340° 19.3
Total: 5 320° 124

4. Measure the Leg Distance

Now you can enter the leg distances on your flight log. Simply look down the Route Page on your GPS and enter in the leg distance figures. Enter the distance on your flight log. Except for the GPS 92, the GPS displays a decimal figure: round up or down to the nearest nautical mile.

Note: If you've got a GPS III Pilot you may need to use the left/right rocker pad function to scroll the displayed field. Be sure to use the Leg Dist display and not the Distance display (which is the cumulative distance at each point).

Now you ought to do a "gross error check". Put a ruler to each line on the chart to check quickly that the distances are correct. The GPS means that you don't need to take time measuring the lines accurately with a ruler, but it's still worth doing a quick check.

5. Measure the True Track

You already have the magnetic track (called the DTK or Course) shown in the GPS, so it's very easy to compute your true track. Simply work out the true track by adjusting the magnetic variation from the magnetic track and enter it into the flight log. At the time of writing the magnetic variation on the example route is around 4 degrees west, so you simply subtract 4 degrees from the magnetic track and enter that in the flight log.

Again you ought to do a "gross error check". Put a protractor to each line on the chart to check quickly that the angles look correct. The GPS means that you don't need to take time measuring the angles accurately with the protractor, but it's still worth doing a quick check.

Tip: You might have spotted from Chapter 8 that you can configure the GPS to display True Track instead of Magnetic Track with auto-magnetic adjustment. This is correct, however its probably not worth setting the GPS to True Track just for the purposes of this section, since most of the time you will want the GPS in Magnetic Track mode.

6. Compute your True Airspeed

Initially you have quite a climb ahead of you, which the plane should manage during the first short leg. For the first leg in the climb use an estimated TAS for the purposes of this exercise of 85Kts.

You will be flying a PA28 with an indicated airspeed (IAS) in the cruise of 100Kts. Imagine once again that looking at Met form 214 you can see that the temperature is 10 degrees at 2000 feet and 4 degrees at 5000 feet.

You are going to fly at flight level 45 over the mountains as this is the correct quadrantal level for the direction you are heading. The temperature will probably be about 5 degrees (by interpolation). The TAS will be about 106Kts.

As explained in the previous chapter, this can be calculated on the GPS but is probably done far more quickly on a wizzwheel.

7. Compute your True Heading

The imaginary Met form 214 indicates a wind of 150 degrees at 15 knots (150/15). Using your wizzwheel again, compute your True Heading, i.e. the heading you will need to steer to achieve the True Track. You will remember that unfortunately this is one really useful function that the Garmin GPS receivers lack.

8. Compute your Magnetic Heading

Adjust the True Heading for Magnetic Heading. In the example the magnetic variation is around 4°W. You need to add 4 degrees.

9 & 10. Compute your Time and Fuel Requirement

Compute the fuel requirement for each leg based on the stage of flight and the published fuel settings for your aircraft.

In the previous chapter you did this computation using a wizzwheel. In this chapter, you will learn how the GPS can help you.

> **Tip:** This next set of instructions is quite complicated and is included for completeness. You will find it quicker and more straightforward to use a wizzwheel (or calculator) for these computations. The author uses a "ballpark" speed and fuel consumption as a "gross error check" for the whole route rather than using the GPS to compute individual legs independently.

Each model is shown separately:

GPS 92

From the Main Menu, select the **E6-B Menu**, then **Trip & Fuel**. You can now compute the fuel on either a route or waypoint basis. By default the page will show your current route (route 1) and the total distance across **ALL** the legs. **Rte 1** should be displayed at the top left (if **Waypoint** is shown, select it, press **ENTER** and use up/down to change it to **Rte** and if necessary edit the numeric field to read **1**).

Next you need to select each leg in turn. Highlight **All** and press **ENTER** and use up/down to select **Leg 1** and press **ENTER** again. Now you can enter the speed from the Flight log for this leg (90kts) and the fuel flow. You may need to refer to the aircraft's POH for the fuel flow. You are in the climb on this first leg so in this aircraft it's likely to be around 12 USG per hour (46L). Write the Time (ETE or Estimated Time En-route) on the flight log and the fuel requirement. Round the times up or down to the nearest minute. You might wish to be conservative and round fuel up. Repeat for the remaining legs. Referring to the POH, you probably will be using around 9 USG (34L) per hour in the cruise.

> **Tip:** Avoid entering a route into Route 0 as the GPS 92 uses Route 0 to store the active route. Any route stored in it will be overwritten when another route is activated.

GPS III Pilot

From the Routes Page, press 🔘 once to activate the context sensitive menu and select **Setup Plan**. Enter the **Speed** and **Fuel Flow** from the flight log for the first leg. You may need to refer to the aircraft's POH for the fuel flow. You are in the climb on this first leg so in this aircraft it's likely to be around 12 USG per hour (46L). You can ignore the depart time and date. When both the figures have been entered, press 🔘 to return to the **Route Plan** page. Use the left/right rocker function to display the **Leg Time** and **Fuel** requirement in turn and note this on the flight log. Round the times up or down to the nearest minute. You might wish to be conservative and round fuel up. Repeat for the remaining legs (because although you can see data for all the legs on the plan they are computed based on the leg 1 data, which is not valid for all the legs). You can refer to the flight log for the speed and, referring to the POH, you probably will be using around 9 USG (34L) per hour in the cruise.

GPSMAP 195

From the Main Menu, select **Trip & Fuel Planning**. You can now compute the fuel on either a **route** or **waypoint** basis. By default the page will show a **Point-to-Point Plan**. You need to look at your route on a waypoint by waypoint basis so you need to press 🔘 once to activate the context sensitive menu and select **Route Planning**. Next ensure that your current Route (01) is displayed (highlight the route number and change it if necessary).

Next you need to select each leg in turn. Highlight **All** and press 🔘 and use up/down to select **Leg 1** and press 🔘 again.

Now you can enter the speed from the Flight log for this leg (90kts) and the fuel flow. You may need to refer to the aircraft's POH for the fuel flow. You are in the climb on this first leg so in this aircraft it's likely to be around 12 USG per hour (46L). You can ignore the depart date and time. Write the Time (**ETE** or Estimated Time En-route) on the flight log and the fuel requirement. Round the times up or down to the nearest minute. You might wish to be conservative and round fuel up. Repeat for the remaining legs. Referring to the POH, you probably will be using around 9 USG (34L) per hour in the cruise.

GPSMAP 295

On the Route Plan page press 🔘 once to activate the context sensitive menu and select **Set Fuel Flow**. Enter the Fuel Flow from the flight log for the first leg - you may need to refer to the aircraft's POH for this. You are in the climb on this first leg so in this aircraft it's likely to be around 12 USG per hour (46L). Now enter the speed for the first leg. Highlight the speed field at the top of the screen and change it to match the time on the flight log (90kts). Now press 🔘 once again to activate the context sensitive menu and select **Change Fields**. Select the **Course** field and press 🔘 and edit it to read Leg Fuel (not just Fuel which is the cumulative fuel display).

Flight Planning an Entire Route | Computing Time & Fuel

Note the Fuel and Leg Time requirement on the flight log. Round the times up or down to the nearest minute. You might wish to be conservative and round fuel up. Repeat entering the Fuel and Speed for the remaining legs (because although you can see data for all the legs on the plan they are computed based on the leg 1 data, which is not valid for all the legs). You can refer to the flight log for the speed and, referring to the POH, you probably will be using around 9 USG (34L) per hour in the cruise. When you have finished, reset the **Leg Fuel** column back to **Course**.

> **Fuel Units:** The GPS doesn't specify fuel units, so you can work in Litres, US Gallons or Imperial Gallons, whichever you are used to. Just make sure you remain consistent!

Remember to add Taxi fuel (normally a couple of gallons/10 litres) and holding/diversion time (normally 30-45 minutes) to get the full requirement.

| 2000' W/V | 150 / 15 | TEMP. | +10 | | | |
| 5000' W/V | 150 / 15 | TEMP. | +4 | | | |

DEPARTURE INFORMATION

FROM / TO	MSA	PL/ALT	TAS	TR (T)	W/V	HDG (T)	HDG (M)	G/S	DIST	TIME	ETA	ATA
KEMBLE / GLOUCESTER	2000	CLIMB	85	343	150/15	345	349	100	14	8		
GLOUCESTER / SHOBDON	2400	4500	106	310	150/15	307	311	120	34	17		
SHOBDON / LLANBEDR	4000	4500	106	308	150/15	305	309	119	57	29		
LLANBEDR / CAERNARFON	4600	DESCENT	106	337	150/15	338	342	121	19	10		

Example flight log. Note: Different GPS databases may cause the results shown to differ by a degree or so.

Flight Planning an Entire Route — **Computing Time & Fuel**

11. Fan Lines ("track guides")

You will be learning how to use the GPS to help correct track errors this time, so you can leave track lines off. If you are still studying for your PPL or recently qualified you should continue using Fan Lines until you are comfortable without them.

12. Track Markings

The author uses 6 minute markers. You should use the marking method you prefer.

Finishing off the Planning...

Now you are ready to go. You've checked the weather and worked out that you are going to talk to Kemble, Gloster, Shobdon, Shawbury, Llanbedr, Valley and Caernarfon. You might want to consider filing a flight plan as this route takes you over some pretty inhospitable territory.

Activating the Route

You now need to activate the route in the GPS:

Activating the Route			
GPS 92	**GPS III Pilot**	**GPSMAP 195**	**GPSMAP 295**
On the Routes Page, ensure that Route 1 is still selected and highlight **Act?** in the bottom right of the screen. Press [Enter]. The Route will automatically be copied into Route 0 and made Active. A new page, the Active Route page will appear in the main page loop.	On the Route Plan page displaying your route, press [Menu] once to activate the context sensitive menu and select **Activate** and press [Enter] to activate it. Your route will now appear in the Active Route page in the main page loop. You can also activate your route by highlighting it in the list of routes on the Routes page and activating it in a similar way.	On the Route Edit page displaying your route, press [Menu] once to activate the context sensitive menu and select **Activate Route** and press [Enter] to activate it. Your route will now appear in the Active Route page in the main page loop. You can also activate your route by highlighting it in the list of routes on the Routes page and activating it in a similar way.	On the Route Plan page displaying your route, press [Menu] once to activate the context sensitive menu and select **Activate** and press [Enter] to activate it. Your route will now appear in the Active Route page in the main page loop. You can also activate your route by highlighting it in the list of routes on the Routes List page and activating it in a similar way.

Flight Planning an Entire Route **Activating the Route**

Activating the Route

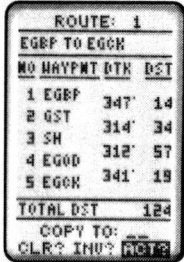

GPS 92

GPS III Pilot

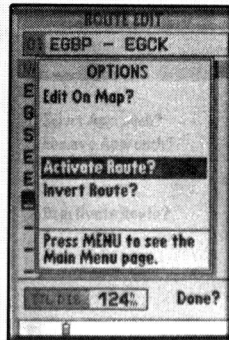

GPSMAP 195

GPSMAP 295

Flying the Route on the Simulator

You are going to be using the Active Route page on this trip and comparing it with your Flight log. The different GPS models can display various useful information on the Active Route page. If you configure the GPS to display the ETA at each point you will always have a reference if an Air Traffic Service asks you for an estimate at a reporting point:

Configuring the Active Route page			
GPS 92	**GPS III Pilot**	**GPSMAP 195**	**GPSMAP 295**
By default the Active Route page displays the Estimated Time Enroute (ETE) and the Distance (DST). To change the ETE to ETA, select ETE and press [Enter] and then use the rocker pad up/down to select **ETA.**	The Active Route page displays the list of waypoints and a single field. You can very quickly use the left/right function of the rocker pad to display ETA (or any other data you wish to see).	By default the Active Route page already displays the distance (DST) and ETA. You can edit these fields by pressing [Menu] once to activate the context sensitive menu, selecting **Change Data Fields**, selecting the field to change, pressing [Enter] and then choosing the data to be displayed.	By default the Active Route page already displays the Course, Distance and the Estimated Time Enroute to the Next waypoint (ETE Next). You can edit these fields by pressing [Menu] once to activate the context sensitive menu, selecting **Change Fields**, selecting the field to change, pressing [Enter] and then choosing the data to be displayed. Try replacing **ETE Next** with **ETA Next.**

Flight Planning an Entire Route   **Flying the Route on the Simulator**

Note: The ETA won't display until you start moving.

Tip: Remember that using ETA is just the Author's recommendation. You can change these fields to match your own preferences.

Kemble to Gloucestershire

Now start the Simulator and initialise the position to Kemble (EGBP). If you've forgotten exactly how to do this, look back at Chapter 9.

For simulation purposes set the speed to your first leg ground speed of 100kts. Set the Altitude to 4500ft (even though in real life it will take a while to climb that high).

Check on the Active Route page that the route is active. The active leg should be to GST. This will be marked (except on the GPS 92) by an arrow on the Active Route page. If the arrow isn't visible or the active leg isn't GST then check that you have correctly initialised the Simulator position to Kemble and that you have entered and activated the route correctly. Try re-initialising the Simulator and activating the route again if you have problems.

Active Route Pages

GPS 92

GPS III Pilot

GPSMAP 195

GPSMAP 295

Write down the departure time from Kemble on the flight log (as indicated on the GPS Navigation page, which has the clock display), and work out the ETA at Gloucestershire. You can compare this with the ETA shown by the GPS on the Active Route page - they should match.

Flight Planning an Entire Route | Kemble to Gloucestershire

Gloucestershire to Shobdon

As you approach GST you can imagine doing the appropriate checks as you did in the previous chapter (HAT checks and/or FREDA checks or equivalent), however this time you don't have to worry about setting up a Goto as the GPS will automatically sequence onto the next waypoint in the route.

Waypoint Check: One important check to do when following a route is to make sure that the GPS has sequenced on to the next waypoint. If it doesn't you will start to see very confusing tracks and distance information! (There should be no problems in Simulator mode unless you have inadvertently started to steer the Simulator manually).

Adjust the Simulator speed to 120Kts to approximate your ground speed as per the flight log. Remember to mark your ATA on the flight log and work out your ETA for the next waypoint. Check the ETA against the one forecast by the GPS.

Winds Aloft: In real life, if the flight log ETA differs markedly from the estimate from the GPS, this is an early indication that winds aloft may not be as forecast (provided the aircraft is being flown accurately). On the Simulator they should match!

Take some time to page through the Main Page loop and become more and more comfortable with the information at your fingertips. Notice on the aeronautical chart that you are flying parallel to airway B39 but on the GPS Map page that there is no sign of it. Similarly as you approach Shobdon there is no sign of either Airway N862 or A25.

Jeppesen Database: The Jeppesen database contains CTA information but NOT airway information.

When you get to Shobdon remember to do your checks as described previously.

Shobdon to Llanbedr

You have quite a long leg from Shobdon to Llanbedr. There should be enough time for you to take a look at configuring the VNav profile and also experimenting with the configuration of the Map and HSI (CDI) page.

Configuring the VNav Profile

Now at last you can look at configuring the vertical navigation profile. Looking at the plate for Caernarfon you will see that it has a slightly unusual circuit at 800ft with overhead joins at 1300ft. Consider descending to a standard overhead join above the destination. You will want to be at 1300 feet AGL shortly before arriving there.

> ⚠️ **Alert:** Several comments have already been made about the accuracy of the altitude reporting. The accuracy is good enough to use VNav alerting as an aide-memoire to your descent, however don't use it as a primary source of altitude information.

Configuring the VNAV Profile			
GPS 92	**GPS III Pilot**	**GPSMAP 195**	**GPSMAP 295**
From the Main Menu, select the **E6-B Menu** and then **Vertical Nav.** The Vnav Waypoint **(Wpt)** will be showing the next waypoint (EGOD in the demonstration case). Edit it to be **EGCK** - notice that the GPS lets you quickly select from the waypoints in your active route. Set the target altitude **(To)** to **1300 ft above waypoint** and aim to be there **By 2nm before** it. Set the Profile to be a descent **(At) 500fpm**, and check that **Vnav** is **On.**	From the Main Menu, select **Vertical Nav.** The **Vnav Waypoint** should be showing your destination waypoint (EGCK in the demonstration case). Edit it if not - notice that the GPS lets you quickly select from the waypoints in your active route. Set the **Target Altitude** to **1300ft above waypoint** and aim to be there **By 2nm before** it. Set the **Profile** to be **500 ft/m downwards**, and check that the alerting **Message** is **On.**	From the Main Menu, select **Vertical Navigation.** The **Vnav Waypoint** should be showing your target destination (EGCK in the demonstration case). Edit it if not - notice that the GPS lets you quickly select from the waypoints in your active route. Set the **Target Altitude** to **1300 ft above waypoint** and aim to be there **By 2nm before** it. Set the **Profile** to be **500 ft/m downwards**, and check that **VNav Messages** is **On.**	From the Main Menu select the **VNAV** tab. The **Vnav Waypoint** should be showing your target destination (EGCK in the demonstration case). Edit it if not - notice that the GPS lets you quickly select from the waypoints in your active route. Set the **Target Altitude** to **1300 ft above waypoint** and aim to be there **By 2nm before** it. Set the **Profile** to be **500 ft/m downwards**, and check that **VNav Messages** is **On.**

VNav Profile Pages

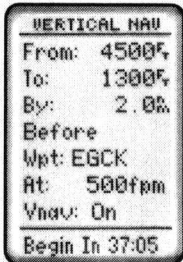

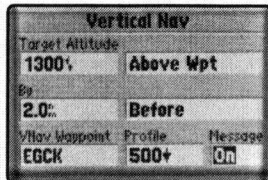

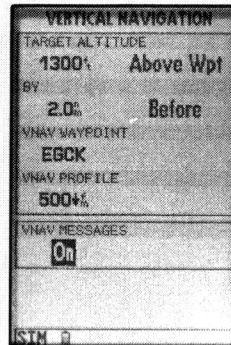

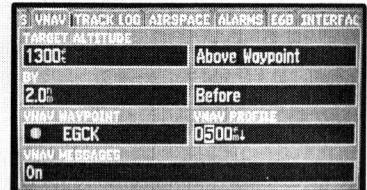

| GPS 92 | GPS III Pilot | GPSMAP 195 | GPSMAP 295 |

Cruise Check

This is quite a long leg. Probably time for an imaginary cruise (FREDA) check. It's also worth checking you can identify where you are visually against the ground and the chart. Don't get lulled into a false sense of security - GPS receivers don't fail often but when they do it tends to be at most inopportune moments!

Configuring the Map Page

The Map page can be configured to display lots of different types of information. It's not a good idea to make these kinds of configuration changes in real flight but doing so during a simulated flight enables you to see your configuration changes in action.

You might want to experiment with different fields on the map screen and the HSI screen (CDI screen on the GPS 92). The different models have quite a diverse set of fields set up by default. Exactly what fields you set up are a matter of personal choice.

The Author's preferences are as follows:

Name of Next Waypoint. The next waypoint isn't always immediately visible. Having it as a field ensures stupid mistakes aren't made and that the GPS is aiming at the same waypoint that you think you are!

Distance to Next Waypoint. Enables distance and (through mental arithmetic) time estimates to be given if requested without changing the page.

Bearing to Next Waypoint. This should match the desired track on the flight log! If it doesn't, left or right drift has taken place. Whilst the GPS feature lots of track error measurements, this measurement gives the best all round feel of track error as well as a bearing direct to the waypoint.

Ground Speed. Useful to compare actual against forecast speed and (through mental arithmetic) time estimates without changing the page.

The GPSMAP 195 and GPSMAP 295 allow multiple configurations of fields including an HSI (on the map page), large size fields and up to 8 fields in total.

Where 8 fields are available the Author recommends using them all, setting the second four as follows:

Track (TRK). The Track "made good" - i.e. the actual track over the ground. If the forecast heading is being accurately steered, the track made good is a direct indicator of the actual drift. If the wind forecast is accurate the Track should match the Desired Track.

Course/Desired Track (DTK). The desired track is readily available on the flight log, but it can be useful to have it readily available on the GPS screen. The track line on the map is the course/desired track.

Estimated Time Enroute (ETE). The estimated time enroute to the next waypoint

Estimated Time of Arrival (ETA). The estimated time of arrival at the next waypoint. Useful in providing estimates if requested by Air Traffic Services. Interestingly the ETA isn't available on the GPSMAP 195.

Configuring the Map page			
GPS 92	**GPS III Pilot**	**GPSMAP 195**	**GPSMAP 295**
The fields on the Map page are fixed. The top left is the bearing to the next waypoint. The bottom left is the actual ground track "made good" (this is NOT the desired track). The top right is the distance. The bottom right is the ground speed.	By default the Map page displays the speed, Time to Next (Estimated Time Enroute), The Distance to Next and a pointer arrow. To change these press [Menu] once to activate the context sensitive menu and select **Change Fields.**	By default the GPSMAP 195 is set up with 4 fields. The Desired Track, Distance, Track and Speed. To change the number and type of fields, press [Menu] once to activate the context sensitive menu and select **Number of Data Fields**. To change the fields press [Menu] once to activate the context sensitive menu and select **Change Data Fields.**	By default the GPSMAP 295 is set up with 2 fields (Distance and Speed) and an HSI. To change the number and type of fields, press [Menu] once to activate the context sensitive menu and select the kind you require. To change the fields press [Menu] once to activate the context sensitive menu and select **Change Fields**.

GPSMAP 295 - Example Fields

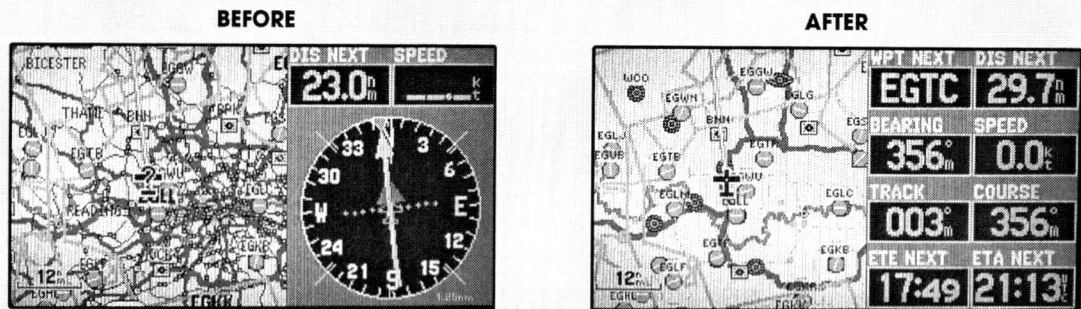

BEFORE

AFTER

Configuring the HSI Page

The HSI page can be configured in a similar way to the map page (the CDI page on the GPS 92 can't be reconfigured).

The Author recommends using similar field settings to the map page so that whan you switch between pages the fields are in the same place. The HSI pointer indicates the course/desired track and the GPS will normally automatically set it to the track to take you from one waypoint to the next. In addition the Course to Steer pointer on the HSI provides the same information as setting one of the fields to CTS. The course to steer is an efficient heading to get you back on track.

Cruise Check

This is quite a long leg. Probably time for another imaginary cruise (FREDA) check. It's also worth checking you can identify where you are visually against the ground and the chart, especially as you are about to start deliberately steering off track!

Manually Steering the Simulator

So far you have been using the Simulator in a mode that automatically follows the route.

Now you can try manually steering the Simulator and see the effect on the Map page and the HSI page.

To steer the Simulator manually, go to the HSI page and use the left/right function of the rocker pad to steer left and right. The HSI will turn to show your new track.

Turn to the map page and see the result. Watch the fields showing the various tracks as they change and start to diverge!

Simulator: Once you start steering manually you are stuck with it until you reset the Simulator. On the GPS III Pilot and GPSMAP 195 you can set the Simulator mode back to Auto Track mid flight on the Simulator set up screen. The GPS 92 and GPSMAP 295 aren't really geared for this and you have to switch of the Simulator and start over.

Discussion: Heading versus Track

It's important to make a distinction between heading and the different definitions of track.

In normal navigation you deal with True Track, Magnetic Track and Magnetic Heading:

- **True Track** is the track actually flown over the ground relative to True North normally measured between two waypoints.

- **Magnetic Track** is True Track corrected for magnetic variation.

- **Magnetic Heading** is the Heading of the aircraft according to the Compass (or a correctly aligned Direction Indicator, and taking into account compass deviation error). In the context of a flight log, the magnetic heading is normally a computed heading based on the desired magnetic track and the wind. If the aircraft steers the magnetic heading and the wind forecast is accurate it should achieve ("make good") the desired track on the ground.

It is important to realise that in the context of the GPS there is no such thing as a heading. The GPS Compass and HSI displays show Track "made good", i.e. the track over the ground NOT the heading of the aircraft. Thus if the aircraft is flying in a strong northerly wind and "laying off" 10 degrees of drift to maintain a track of due East, the real world compass, DI or HSI will read 080° whilst the GPS Compass and HSI will read 090°.

Flight Planning an Entire Route   **Heading Versus Track**

Aircraft HSI Reading 080

GPS HSI reading 090

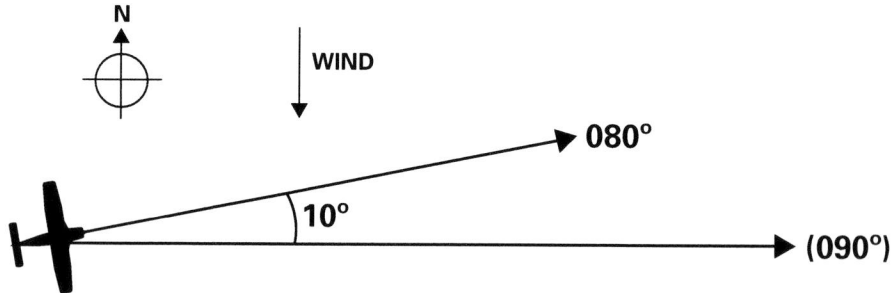

Diagram illustrating the different indications between a conventional aircraft HSI and the HSI of a GPS. The aircraft is tracking 090 but heading 080. The GPS can't tell which direction the aircraft is heading, only the direction it is tracking.

The GPS can show:

- **Track "made good".** The actual track over the ground.

- **Desired Track** (also known as Course). The desired track between two waypoints.

- **Bearing.** The track from the current position to a waypoint. This will be the same as the desired track, ONLY if you are on the desired track.

- **Cross track error**. The distance at right angles from the desired track to the current position (i.e. how far off track you are).

- **Course to Steer.** Recommended track from the current position in order to regain the desired track (computed by the GPS taking into account the cross-track error bearing and desired track).

Fact: At no point does the GPS show your heading!

Llanbedr to Caernarfon

By now you are probably approaching Llanbedr. Remember to do all the standard checks. How did the ETA on the flight log work out?

Compute the ETA at Caernarfon. How does it match the one the GPS forecasts?

You probably need to steer slightly East of track to stay out of Danger area 202. You can pick up the track on the other side of the bay.

Fairly soon you will receive the VNav message indicating that you need to be starting your descent. There is a critical MSA here and the track takes you nearly straight through a 2050ft mast. You may want to steer slightly west of track to keep down the valley. Make sure you have the mast visually outside the (imaginary) cockpit window.

Your position relative to the GPS track should make it far easier to look in exactly the right direction to spot it.

Now you can call up Caernarfon and join and land. Remember that you have no more use for the GPS once the airfield is in sight!

Flying the Route for Real

When you are confident about planning and flying routes on the simulator you can try flying a route for real. Here are the things you should look out for again:

Alert: The first time you use the GPS for real go up with a safety pilot, or better still sit in the right hand seat and do the navigating before progressing to the left hand seat.

Checklist: It's a good idea to switch on your GPS at the same time you switch on your other avionics. This is normally shortly after you have started the engine. Remember to check the database validity and allow sufficient time for the receiver to acquire satellites.

Fly the Heading not the Line: Fly the heading on your flight log, not the line on the GPS. If the wind forecast is correct and you fly accurately this will take you down your desired track. How to correct for track errors is explained in Chapter 16.

Checklist: When you do your cruise checks is a good time to check the GPS is functioning properly. Page to the Satellite Status screen. Do you have good coverage? Is the Battery OK (or are you still on external power)? Does the waypoint the GPS is aiming at match the one you are really aiming at?

Flight Planning an Entire Route | **Flying the Route for Real**

Tip: When you are airborne things are very different from on the ground and you may find you temporarily forget some of your GPS training! The following tip works anywhere to get you back to the Satellite Status page: Just keep on pressing (QUIT). No matter where you are, you will soon be back to the Satellite Status page.

Alert: Once you have the destination airfield in sight, STOP fiddling around with the GPS and make sure you have done your airfield approach checks and in due course your pre-landing checks.

Checklist: As you shut the aircraft down you will probably switch the avionics off. Now is a good time to switch off the GPS.

Two views of the mast on the approach to Caernafon in inclement weather.

Introduction

To get the most out of your GPS it's important that you are fully conversant with waypoints and their uses. You've used waypoints a lot already in the previous Chapters. This short Chapter looks briefly at waypoint information stored in the GPS.

Waypoint Types

The Jeppesen database stores information about airfields and navaids. With the exception of the GPS 92, the receivers also have "base maps" which include information on roads, railways, rivers, towns and cities.

The GPSMAP 295 contains even more data including more cities and features than other models, motorway intersections and points of interest. With the addition of a MapSource cartridge and CD the GPSMAP 295 can even store information down to street address level.

To get the most out of your GPS it's important that you are fully conversant with waypoints and their uses.

Airport Data - Airport data is displayed in three main groupings:

- Location information. This includes the Facility and City names as well as Latitude and Longitude, Elevation and Fuel information.

- Runway Information. This includes runway directions, lengths, surface and lighting information

- Communication facilities. These include frequencies of all listed frequencies for the airport. Where multiple frequencies of the same type appear, the one listed first is usually the standard frequency listed in flight guides.

Navaid Data

- Navaid data is included for NDBs, VORs, (including TACANs), and Intersections.

- NDB information. This includes the Facility and City name, the frequency and location.

- VOR Information. This includes the Facility and City name, the frequency, whether a DME is co-located, and location.

- Intersection Information. This includes location and relative bearing and distance from a nearby radio aid.

Viewing Waypoints

Waypoint information can be viewed specifically or in conjunction with another operation such as selecting a waypoint on a map or reviewing a waypoint as part of route entry or editing.

To view waypoint information proceed as follows:

Viewing Waypoint Information			
GPS 92	GPS III Pilot	GPSMAP 195	GPSMAP 295
Press the [Wpt] button.

Using the cursor highlight the waypoint type at the top left of the screen and press [Enter].

You can choose to view airports **(APT)**, intersections **(INT)**, NDBs, VORs or user waypoints **(USR)**, by using the up/down rocker pad function.

If you select the waypoint name, facility or city you can use the rocker pad to select the waypoint displayed.

When viewing airports, you can use the options at the bottom of the screen to display the other pages of information.

If you select the user waypoint type you can rename, delete or edit the waypoint or create a new one. | From the Main Menu, select **Waypoints.**

You can choose to view airports **(APT)** (all three pages of information), intersections **(INT)**, NDBs, VORs or user waypoints **(USR)**, by using the left/right rocker pad function to select the appropriate menu tab.

If you select the waypoint name, facility or city you can use the rocker pad to select the waypoint displayed.

If you select the **User Wpt** tab you can rename, delete or edit the waypoint or create a new one.

If you select the **User List** tab you can delete all the waypoints. | Press the [Wpt] button.

Using the cursor highlight the waypoint type at the top left of the screen and press [Enter].

You can choose to view airports **(APT)**, intersections **(INT)**, NDBs, VORs or user waypoints **(USR)**, by using the up/down rocker pad function.

If you select the waypoint name, facility or city you can use the rocker pad to select the waypoint displayed.

When viewing airports, you can use the options at the bottom of the screen to display the other pages of information.

If you select the user waypoint type you can rename, delete or edit the waypoint or create a new one. | Press the [Goto/WPT] button.

You can choose to view aviation (airport location data, intersections, NDBs and VORs), Runway information, Comm Frequencies, Approaches, User waypoints, Cities, Motorway Exits, Points of Interest (POI) and, if you have a data card installed, even Addresses and Crossroad intersections by using the left/right rocker pad function to select the appropriate menu tab.

If you select the waypoint name, facility or city you can use the rocker pad to select the waypoint displayed.

If you select the **User** tab you can rename, delete or edit the user waypoint or create a new one.

By activating the context sensitive menu you can manage the entire list and set up a reference waypoint or a "Favourites" list. |

Spend some time familiarising yourself with finding your way around the waypoint information screens. Getting at waypoint data quickly and efficiently, especially frequencies, can be very useful.

Introduction

In this Chapter you will learn about creating and using user-defined waypoints. You'll also learn how to enter different waypoint types, how to insert waypoints in the middle of a route and, if you've got a GPSMAP 195 or 295, how to edit routes using the map.

Sometimes you will want to route via major landmarks or towns rather than via aeronautical features. Depending on which model GPS you have, you will find greater or lesser support for waypoints, so sooner or later this means having to create a user-defined waypoint.

The next leg of your journey will take you from Caernarfon to Manchester Barton. You can use either a Southern England and Wales or Northern England 1:500000 chart for this, though the Northern England chart is probably better.

Creating a User-defined Waypoint

There are four ways of creating a user-defined waypoint:

- Entering its latitude and longitude
- Choosing a town from the base map (not GPS 92)
- Entering a distance and relative bearing from another waypoint
- Marking it on the map (not GPS 92)

Whichever method you use, its worth cross checking it with another method and also verifying the bearing and distance to the waypoint when you first use it in a route. Not all the GPS receivers in the range support all the methods.

The example route is along the Menai Straits via the Menai Bridge to Llandudno to keep clear of the high ground, then across to Ashcroft airfield at the southern side of the low level route between Manchester and Liverpool, up the low level route to the junction of the M62 and M6 and then direct to Barton.

Using Latitude and Longitude

Create the first waypoint by specifying its latitude and longitude. You can measure the position on a chart or use a book of GPS coordinates to give you the position. You can measure latitude on a 1:500000 chart quite easily as each nautical mile is one minute of latitude. For the longitude its best to line up a ruler at right angles to intersect the nearest grid mark and read off the map.

Using Latitude and Longitude

GPS 92	GPS III Pilot	GPSMAP 195	GPSMAP 295
From the Main Menu select **User Wpt List.** Press the [WPT] button once and select **New.** This will bring up a blank user waypoint screen. Move the cursor to the top (name) field and enter in a name of up to 6 characters. Enter MENAI for the example. Press [Enter] when you have finished. Next you need to enter the latitude and longitude. By default the GPS will put in the current position. Edit the position to **N53°13.500' W004°10.000'** and press [Enter]. The GPS will enter the time the waypoint was created. You can go up and edit this if you wish (for example to spell Menai Bridge in full). Finally select **Done** and press [Enter].	From the Main Menu select **Waypoints.** Use the rocker pad left/right to select the **User List** tab. Press [Menu] once to activate the context sensitive menu and select **New Waypoint.** A new waypoint will be created and named using a number (001 if it's the first one). Move the cursor to the top (name) field and enter in a name of up to 6 characters. Enter MENAI for the example. Press [Enter] when you have finished. Next you need to enter the latitude and longitude. By default the GPS will put in the current position. Edit the position to **N 53°13.500' W004°10.000'** and press [Enter]. The GPS will enter the time the waypoint was created. You can go up and edit this if you wish (for example to spell Menai Bridge in full). Finally select **Done** and press [Enter].	Press the [WPT] Button. Press the [Menu] button and select **Create Waypoint.** A new waypoint will be created and named using a number (001 if it's the first one). Move the cursor to the top (name) field and enter in a name of up to 6 characters. Enter MENAI for the example. Press [Enter] when you have finished. Next you need to enter the latitude and longitude. By default the GPS will put in the current position. Edit the position to **N 53°13.500' W004°10.000'** and press [Enter]. The GPS will enter the time the waypoint was created. You can go up and edit this if you wish (for example to spell Menai Bridge in full). Finally select **Done** and press [Enter].	Press the [Goto/WPT] Button. Select the **User** tab and highlight **New** on screen and press [Enter]. A new waypoint will be created and named using a number (001 if it's the first one). Move the cursor to the top (name) field and enter in a name of up to 10 characters. Enter MENAI for the example. Press [Enter] when you have finished. Next you need to enter the latitude and longitude. By default the GPS will put in the current position. Edit the position to **N 53°13.500' W004°10.000'** and press [Enter]. Finally select **OK** and press [Enter].

User Defined Waypoints — **Using Latitude & Longitude**

Creating MENAI

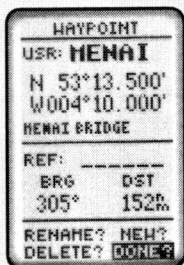

GPS 92

GPS III Pilot

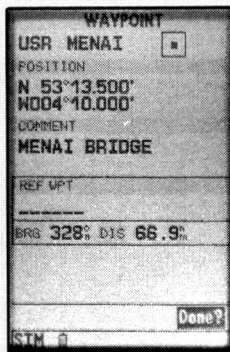

GPSMAP 195

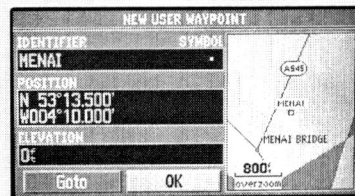

GPSMAP 295

Fact: The GPSMAP 295 already has thousands of features, towns and cities in its database (in fact it already has the Menai Bridge). In fact, you could have found the Menai Bridge in the built-in database by selecting the City tab, however learning to create your own waypoint is well worth while.

Using a Town or City (not GPS 92)

Quite often you may want to use a town or city as a waypoint. In the example, you need to use Llandudno on the North Wales Coast.

Using a Town or City

GPS 92	GPS III Pilot	GPSMAP 195	GPSMAP 295
Unfortunately you can't select a town directly on the map as there is no base map data to look at. You will have to enter the waypoint directly using latitude and longitude as described earlier. For reference it is at about N 53°19.0' W003°48.0'.	On the Map page, scroll and pan the map zooming out and in as required until you find Llandudno on the North Wales Coast. Select it using the pointer on screen. Press [Enter] to create a new waypoint. It automatically names the waypoint with a six letter abbreviation (in this case LNDDNO). Select **Done** and press [Enter].	On the Map page, scroll and pan the map zooming out and in as required until you find Llandudno on the North Wales Coast. Select it using the pointer on screen. Press [Enter] to create a new waypoint. Choose **User Waypoint** as the Waypoint type and edit the name (in this case LNDDNO). Select **Done** and press [Enter].	On the Map page, scroll and pan the map zooming out and in as required until you find Llandudno on the North Wales Coast. Select it using the pointer on screen. Press and hold [Enter] to create a new waypoint. Even though you have up to 10 characters, name it LNDDNO to keep it the same as the other examples. Select **OK** and press [Enter].

User Defined Waypoints / **Using a Town or City**

Fact: on the GPSMAP 295 you could have found the Llandudno in the built-in database by selecting the City tab, however learning to create your own waypoint is well worth while.

Tip: If you are going to use PC flight planning software, for compatibility keep the user waypoint names to six characters or less and avoid using spaces - this is especially relevant to the GPSMAP 295.

Creating LNDDNO

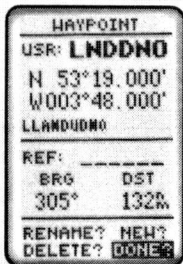

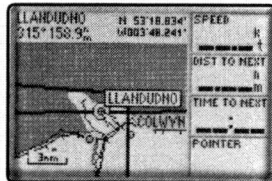

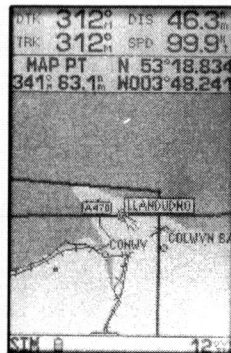

| GPS 92 | GPS III Pilot | GPSMAP 195 | GPSMAP 295 |

Using Distance and Bearing

Now let's enter Ashcroft. This time instead of using the latitude and longitude specify the location as a distance and relative bearing from a waypoint already in the Jeppesen database. Looking at a "plate" for Ashcroft you can see it is 33 miles on the 286 radial from the Trent VOR (Pooley's and Jeppesen vary - your flight guide may proffer alternative VORs). For reference, Ashcroft is at about N 53°10' W002°34'.

Jeppesen Database: You might be wondering why you have to enter Ashcroft as it is an airfield. The Jeppesen database generally has all licensed airfields with an ATZ. It also generally has most airfields that have a four letter ICAO identifier. If you fly to farm strips and small unlicensed airfields, you may find they are not in the Jeppesen database.

In this example you will create the waypoints differently, so you can learn how to edit them at the same time:

| User Defined Waypoints | Using Distance and Bearing |

Using a Distance and Bearing

GPS 92	GPS III Pilot	GPSMAP 195	GPSMAP 295
Press the [Wpt] button twice. This creates a waypoint at the current location. Move the cursor to the top (name) field and enter in a name of up to 6 characters. Enter ASHCFT. Press [Enter] when you have finished, then highlight **Save** and press [Enter] to save it. Next you need to edit the position. From the Main Menu select **User Wpt List** and select **ASHCFT** from the list. Select the **Ref** field and enter in **TNT**. Now enter in the bearing (286) and Distance (33). The latitude and longitude should change to a reading very close to the latitude and longitude shown on your plate. Finally select **Done** and press [Enter].	Press <u>and hold</u> the [Enter] button to "mark" (create) a new waypoint. This creates a waypoint at the current location. Move the cursor to the top (name) field and enter in a name of up to 6 characters. Enter ASHCFT. Next you need to edit the position. Select the **Ref** field and enter in **TNT**. Now enter in the bearing (286) and Distance (33). The latitude and longitude should change to a reading very close to the latitude and longitude shown on your plate. The GPS III Pilot lets you change the symbol for the waypoint: Highlight the square blob to the right of the name and press [Enter]. Now select a suitable symbol such as the circle representing a "soft field" and press [Enter] again. Finally select **Done** and press [Enter].	Press the [Wpt] button twice. This creates a waypoint at the current location. Move the cursor to the top (name) field and enter in a name of up to 6 characters. Enter ASHCFT. Press [Enter] when you have finished, then highlight **Done** and press [Enter] to save it. Next edit the position. Press the [WPT] Button. Press the [Menu] button and **Show User List** and select **ASHCFT** from the list. Select the **Ref Wpt** field and enter in TNT. Now enter in the bearing (286) and Distance (33). The latitude and longitude should change to a reading very close to the latitude and longitude shown on your plate. The GPSMAP 195 lets you change the symbol for the waypoint: Highlight the square blob to the right of the name and press [Enter]. Now select a suitable symbol such as the circle representing a "soft field" and press [Enter] again. Finally select **Next** and press [Enter].	Setting up a new waypoint on the GPSMAP 295 is considerably different from other models. First press [goto/WPT] and check that the Aviation tab is selected. Edit the ICAO code to read TNT and press [Enter]. Now press [menu] once to activate the context sensitive menu and select **Reference Waypoint**. Now enter in the bearing (286) and Distance (33) and select **Create Waypoint**. This creates a waypoint at the defined location. Move the cursor to the identifier field and enter in a name of up to 10 characters. Enter ASHCFT to keep compatible with the other examples. The GPSMAP 295 lets you change the symbol for the waypoint. Highlight the square blob to the right of the name and press [Enter]. Now select a suitable symbol such as the circle representing a "soft field" and press [Enter] again. You can also enter the **Elevation** (150ft from the plate). Finally select **OK** and press [Enter].

Creating ASHCFT

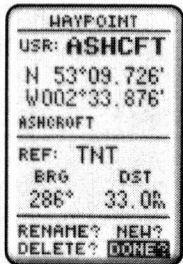

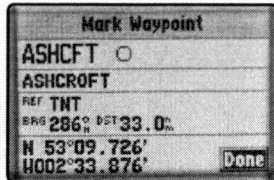

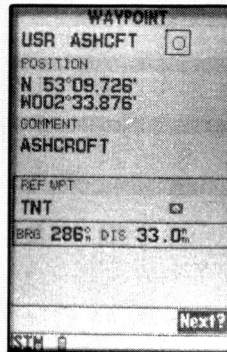

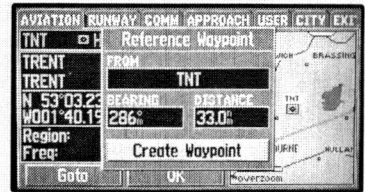

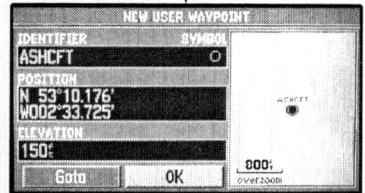

GPS 92	GPS III Pilot	GPSMAP 195	GPSMAP 295

Marking a Waypoint on the Map (Not GPS 92)

Now you need to enter the M6/M62 intersection. Sometimes it can be quicker to find the location on the base map containing roads, railways, rivers and towns than to measure and enter in the co-ordinates.

This works in a similar way to choosing a town, as before, except here you can select any point you like on the map.

Marking a Waypoint on the Map

GPS 92	GPS III Pilot	GPSMAP 195	GPSMAP 295
Unfortunately you can't enter a position directly on the map as there is no base map data to look at. You will have to enter the waypoint directly using latitude and longitude as described earlier. For reference it is at about N 53°25.5 W002°33.5'.	On the Map page, scroll and pan the map zooming out and in as required until you find the junction of the M6 with the M62. It is just north of Warrington. Press [Enter] to create a new waypoint. Edit the name to be M6M62 and select **Done** on the screen.	On the Map page, scroll and pan the map zooming out and in as required until you find the junction of the M6 with the M62. It is just north of Warrington. Press [Enter] to create a new waypoint and choose **User Waypoint** as the waypoint type. Edit the name to be M6M62 and select **Done** on the screen.	On the Map page, scroll and pan the map zooming out and in as required until you find the junction of the M6 with the M62. It is just north of Warrington. Press and Hold [Enter] to create a new waypoint. Edit the name to be M6M62 and select **OK** on the screen.

Creating M6M62

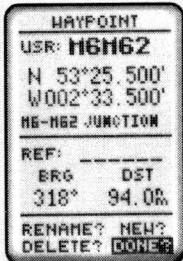

GPS 92

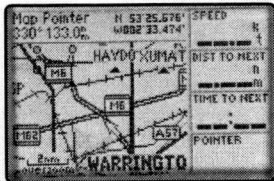

GPS III Pilot

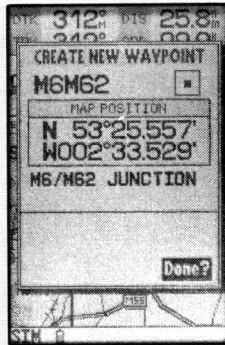

GPSMAP 195

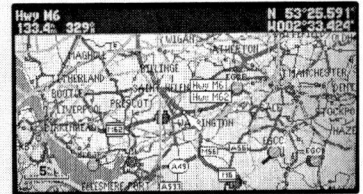

GPSMAP 295

Fact: The GPSMAP 295 actually has motorway intersections in its database. In fact you might have noticed it identify this one as Exit 10/21A of the M6/M62! For the purposes of this lesson go ahead and create the waypoint anyway.

Fact: You can actually create new waypoints at any time when the map is displayed. You can display the map even if you are part way through entering a route by pressing MENU on the Route planning page and selecting Show Map to do the editing.

Entering the Route

Create a new route and enter the first few waypoints: EGCK, LNDDNO and ASHCFT.

Fact: Notice that except for the GPSMAP 295 the GPS lets you enter the waypoints directly into the route. The GPSMAP 295 needs you to select the waypoint type by selecting the appropriate tab: Aviation for all airfields, VORs, NDBs and Intersections, User for User waypoints, City for towns and cities and certain other features.

Inserting a Waypoint

Eagle eyed readers will have spotted that the MENAI waypoint was missed out by accident. If you miss out a waypoint you can insert it without having to rewrite the route from scratch:

Inserting a Waypoint			
GPS 92	**GPS III Pilot**	**GPSMAP 195**	**GPSMAP 295**
Highlight the waypoint after the forgotten waypoint (LNDDNO).	Highlight the waypoint after the forgotten waypoint (LNDDNO).	Highlight the waypoint after the forgotten waypoint (LNDDNO).	Highlight the waypoint after the forgotten waypoint (LNDDNO).
Press [Enter] to activate a context sensitive menu and select **Insert.**	Press [Menu] to activate the context sensitive menu and select **Insert.**	Press [Enter] to activate a context sensitive menu and select **Insert.**	Press [Enter] to activate a context sensitive menu and select **Insert.**
Edit the forgotten waypoint into the space: MENAI.	Edit the forgotten waypoint into the space: MENAI.	Edit the forgotten waypoint into the space: MENAI.	Edit the forgotten waypoint into the space: MENAI.

Fact: You can **Review** (look at the details), **Remove** (Delete) or **Change** waypoints in a similar manner.

Editing the Route on the Map (not GPS 92)

Often it can be quicker to enter consecutive short legs directly onto the map instead of as text. Enter the last two waypoints like this:

Editing the Route on the Map			
GPS 92	GPS III Pilot	GPSMAP 195	GPSMAP 295
Unfortunately the GPS 92 doesn't support this feature. You will need to enter M6M62 and EGCB as above.	Press [Menu] once to activate the context sensitive menu. Select **Show Map.** The first waypoint will be shown. You will need to scroll and pan the map and locate the last waypoint (ASHCFT). Press [Menu] to activate a context sensitive menu. To add the next waypoint select **Insert.** INS appears by the pointer. You can now pan and scroll the map zooming out and in if necessary to select your next waypoint. Scroll northwards and select your M6M62 waypoint (press [Enter]) and then North East to Manchester Barton (EGCB). Press [Menu] and select **Show Text** to return to the Route Edit page.	Press [Menu] once to activate the context sensitive menu. Select **Edit on Map**. The first waypoint will be shown. You will need to scroll and pan the map and locate the last waypoint (ASHCFT). Press [Enter] to activate a context sensitive menu. To add the next waypoint select **Insert**. INS appears by the pointer. You can now pan and scroll the map zooming out and in if necessary to select your next waypoint. Scroll northwards and select your M6M62 waypoint (press [Enter]) and then North East to Manchester Barton (EGCB). Press [Menu] and select **Edit as Text** to return to the Route Edit page.	Select the waypoint space after ASHCFT and press [Menu] once to activate the context sensitive menu. Select **Show Map**. The last waypoint will be shown. Press [Enter] to activate a context sensitive menu. To add the next waypoint, select **Add Turns**. INS appears by the pointer. You can now pan and scroll the map zooming out and in if necessary to select your next waypoint. Scroll northwards and select your M6M62 waypoint (press [Enter]) and then North East to Manchester Barton (EGCB). Press [Menu] and select **Edit as Text** to return to the Route Plan page.

Tip: If you have a GPS III Pilot, GPSMAP 195 or GPSMAP 295 you can actually edit the entire route on the map, adding user defined waypoints as you go. Even if you do this you should still plot the route on an aeronautical chart first.

Flying the Route

Now you can fly the route using the Simulator or in real life using the techniques you learned in earlier Chapters.

The Menai Bridge

Introduction

In this Chapter you'll learn how to detect when you are off track using the GPS and various ways to correct it. You will also learn how to compute the actual "wind aloft".

In conventional VFR Navigation (Deduced or Dead Reckoning) pilots are taught various methods of detecting an off-course situation and various ways to correct it. This is where a GPS really comes into its own. You can very rapidly detect errors in your track and very easily correct them.

Recapping Traditional Methods

Most traditional methods of detecting track error and correcting it are based around using the "one in sixty" rule efficiently. Broadly speaking this says that if an aircraft flies one degree off course for a distance of sixty miles it will be one mile off track.

$$\text{Distance off track (nm)} = \frac{\text{track error (degrees) x distance travelled (nm)}}{60}$$

$$\text{Track Error (degrees)} = \frac{\text{distance off track (nm) x 60}}{\text{distance travelled (nm)}}$$

Fact: Most GPS receivers call the Distance off track the "cross track error". The GPS III Pilot differs and calls this "off course".

You can typically use the one in sixty rule and varying degrees of mental agility (or a wizzwheel) to make a number of assertions, for example:

- If you realise you have been travelling 5 degrees off track for nearly 20 minutes you can expect to be roughly 2.5 miles off track (A light aircraft does about 30 miles in just under 20 minutes in still air so you will be about half a mile off track per degree).

- If you find you are about 2 miles off track after a 30nm leg, your track error is about 4 degrees. (2x60/30). This may cause you to realise you have been mis-steering, or perhaps that the wind aloft is different from forecast.

Your track markers can help you assess the track errors by giving you time or distance information.

How did you get off course?

It's important to realise how you got off course, as this determines your actions to regain and stay on track. These are likely reasons (which may be in combination):

- Mis-steering - the intended heading has not been flown.

- Mis-reading DI - the direction indicator has not been regularly aligned with the Compass (and perhaps not corrected for any compass deviation).

- Winds aloft not as forecast - the wind is not as forecast resulting in an over or under-correction between the desired track and the heading required to "make good" this track.

Correcting Drift

You will probably have been taught to correct drift by assessing the track error and then making an appropriate adjustment. Fan lines (track guides) can help you gauge the likely error and help with corrections.

Parallelling the Desired Track

- Adjusting heading by track error will enable you to fly parallel to the original track.

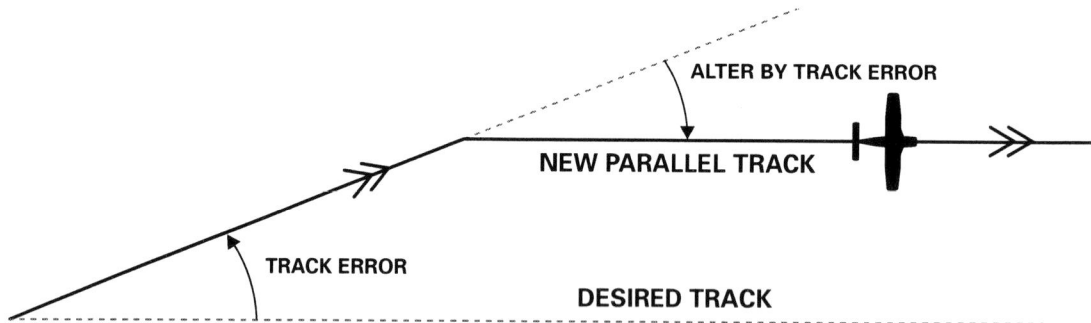

Re-acquiring your Track at the next Waypoint

- You can adjust the heading to re-acquire your route at the next waypoint. To do this you use the one in sixty rule again and adjust your heading by the track error (to cease further divergence) plus an additional amount to give you a closing angle:

$$\text{Closing angle (degrees)} = \frac{\text{Distance off track (nm)} \times 60}{\text{distance to travel (nm)}}$$

Correcting Drift and Track Error　　**Parallelling Track**

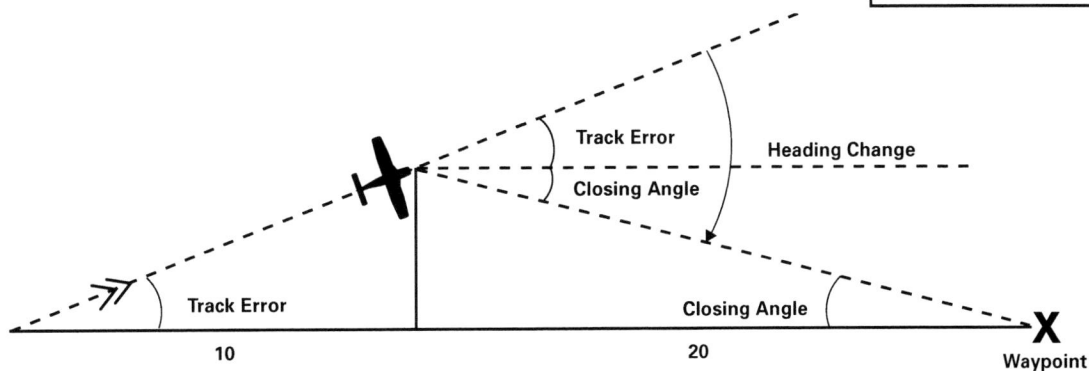

Track Error

Closing Angle

Heading Change

Track Error

Closing Angle

10

20

X

Waypoint

Re-acquiring your Track before the next Waypoint

You can adjust the heading to regain your track before the next waypoint - assuming you haven't yet reached the halfway point on the leg, you can adjust your heading by twice the track error for the same time (and distance) as you diverged and then adjust to your intended heading taking account of the track error. If you are closer then you may need to use a bigger closing angle.

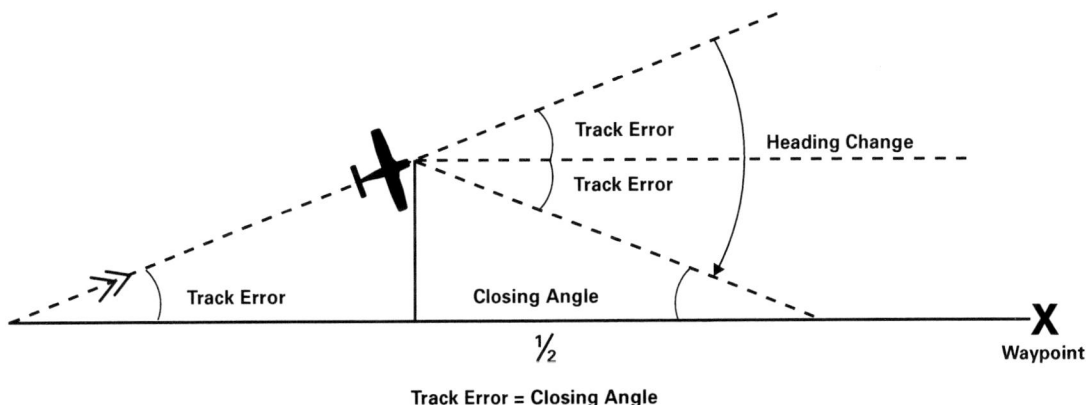

Track Error

Track Error

Heading Change

Track Error

Closing Angle

½

X

Waypoint

Track Error = Closing Angle

If you have been mis-steering or the DI is not aligned correctly then you should take into account when getting back on course that to parallel your original track it may be enough to simply steer the correct heading before adding in a closing angle rather than compensating for track error and the closing angle.

If the wind is not as forecast you need to compensate for your track error and the closing angle.

Correcting Drift and Track Error

Regaining Track

Course Correction using GPS

GPS makes detecting and correcting track error very much simpler and does away with most if not all the mental arithmetic.

The GPS will tell you immediately if you are off track and, if you configure it properly, how much by and how to get back on track.

It's important when using the GPS to still fly the heading forecast by your planning. If you try to "fly the line" you are likely to spend a lot more time with your head in the cockpit than you ought to. Instead bring the GPS occasionally into your instrument scan. When you detect an off course situation, you can apply appropriate corrections to get back on course.

If you over correct you will be able to detect this and make minor corrections back again.

Bracketing the track

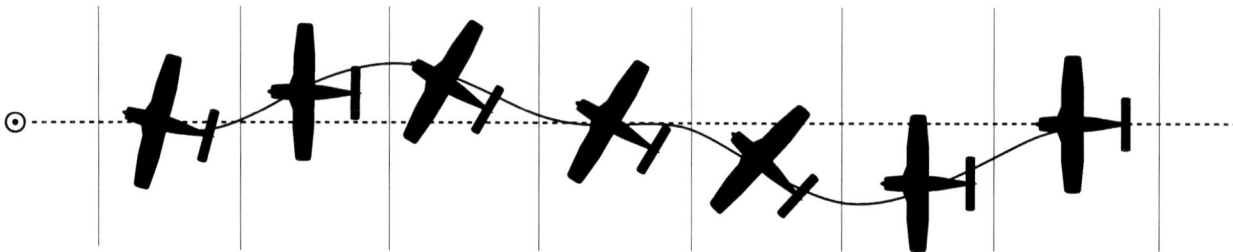

Fact: This is known as "bracketing the track" and is similar to the technique used by instrument pilots tracking towards VORs and NDBs. The technique is most often used when the drift due to wind is initially unknown. With a GPS you can quickly work out the actual drift as you will see later.

Recap: Heading versus Track

Its worth recapping the discussion in Chapter 13 and making the distinction between heading and the different definitions of track.

In normal navigation you deal with True Track, Magnetic Track and Magnetic Heading:

- **True Track** is the track actually flown over the ground relative to True North normally measured between two waypoints.

Correcting Drift and Track Error | **Bracketing the Track**

🌑 **Magnetic Track** is True Track corrected for magnetic variation.

🌑 **Magnetic Heading** is the Heading of the aircraft according to the Compass (or a correctly aligned Direction Indicator, and taking into account compass deviation error). In the context of a flight log, the magnetic heading is normally a computed heading based on the desired magnetic track and the wind. If the aircraft steers the magnetic heading and the wind forecast is accurate it should achieve ("make good") the desired track on the ground.

It is important to realise that in the context of the GPS there is no such thing as a heading. The GPS Compass and HSI displays show Track "made good", i.e. the track over the ground NOT the heading of the aircraft. Thus if the aircraft is flying in a strong northerly wind and "laying off" 10 degrees of drift to achieve a track of due East, the real world compass, DI or HSI will read 080°, whilst the GPS Compass and HSI will read 090°

Aircraft HSI Reading 080   GPS HSI reading 090

The GPS can show:

🌑 **Track "made good".** The actual track over the ground.

🌑 **Desired Track** (also known as Course). The desired track between two waypoints.

🌑 **Bearing.** The track from the current position to a waypoint. This will be the same as the desired track ONLY if you are on the desired track.

🌑 **Cross track error.** The distance at right angles from the desired track to the current position (i.e. how far off track you are).

🌑 **Course to Steer.** Recommended track from the current position in order to regain the desired track (computed by the GPS taking into account the cross-track error bearing and desired track).

Correcting Drift and Track Error   **Heading Versus Track**

Fact: At no point does the GPS show your heading!

Detecting Track Error using the GPS Displays

All the GPS models make it very easy to detect the direction and magnitude of track error. There are some instances where you might wish to improve the information displayed by reconfiguring the GPS as described earlier in Chapter 13.

Fact: The Section that follows talks about course deviation indicators (CDIs). CDIs will be familiar to instrument pilots as they are similar to the VOR display. A CDI is a visual representation of course deviation. A bar moves indicating where the desired track is relative to your current position. If it moves left you are right of track and if it moves right you are left of track. It is known as a "fly to" indicator as, if you fly towards it, you will regain your track.

Fact: Some GPS receivers can be configured to have an arrow pointer that points towards the waypoint. You may recognise its behaviour as similar to an ADF (the instrument which points to an NDB).

Tip: You might want to enable a Goto or a Route on the simulator to see the information described over the following few pages - with no active Goto or Route most of the information will be blank or not appear.

GPS 92 Map Page

The Map page shows the Bearing to waypoint in the top left and Track ("made good") at the bottom left. If you are on course the Bearing and the Track will be the same and equal the Desired Track on your flight log. If the Track does not match then you are heading off course and soon the bearing to the waypoint will change accordingly. Once the errors become significant (and depending on the zoom setting) the plane will diverge from the line on the map.

GPS 92 Navigation (CDI) Page

The Navigation (CDI) page shows the bearing (BRG) and track (TRK) and also the cross track error (XTK). There is also an arrow which points towards the waypoint and a course deviation indicator that visually shows the cross track error and the direction. If you are on course the Bearing and the Track will be the same and equal the Desired Track on your flight log. The CDI will be centred with the arrow pointing ahead and the cross track error will be close to zero. If the Track does not match then you are heading off course and soon the bearing to the waypoint will change accordingly, the CDI marker will start to move left or right (indicating you are right or left of the desired track) and the cross track error will increase. You can change the CDI scale on the Nav Units page of the Setup Menu.

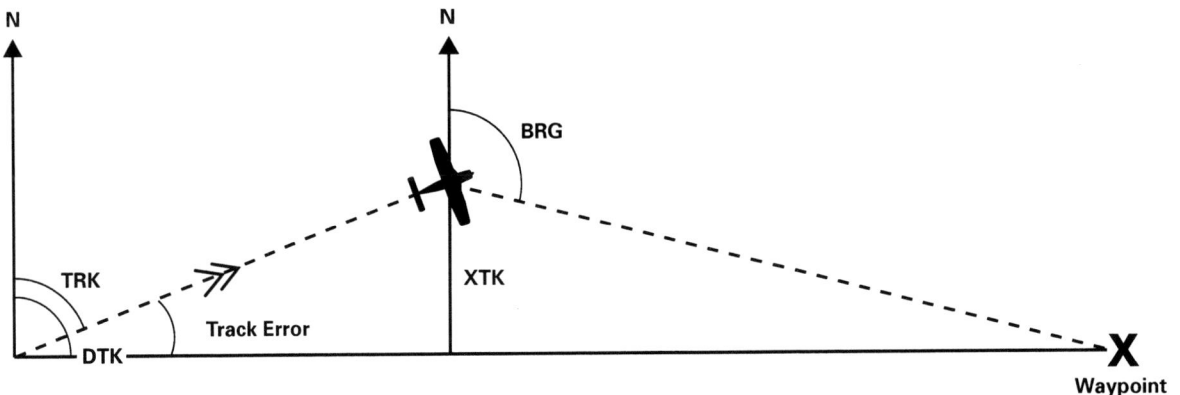

Correcting Drift and Track Error   **Detecting Track Error**

Fact: The Horizontal Situation Indicator (HSI) will be familiar to most instrument pilots and is like a Course Deviation Indicator (DI) superimposed on a Direction Indicator (DI). The DI's compass "card" indicates the track "made good" (not the aircraft's heading which may be different). The main CDI arrow-head shows the desired track. The centre part of the arrow is called the D bar and moves left or right when off course indicating the distance off course. The arrow head near the centre of the HSI indicates if you are travelling "to" or "from" the waypoint (if you are navigating a route it will normally automatically be pointing ahead "to" the next waypoint, however if you have executed a Goto or have overshot the last waypoint in a route, it will point "back" once you pass the waypoint). The "bug" is the pointer around the edge of the compass. By default it recommends a course to steer (CTS).

Correcting Drift and Track Error

Detecting Track Error

GPS III Pilot Map Page

The Map page shows a pointer by default. This points straight if you are tracking directly to the waypoint (regardless of whether you are off track or not). Once the errors become significant (and depending on the zoom setting) the plane will diverge from the line on the map. The Author recommends reconfiguring the GPS to replace the pointer field with Bearing which is much more useful.

GPS III Pilot HSI Page

The HSI page shows a lot of useful visual information as described in the previous panel. If you are on course the CDI will be centred and the DI track indication will match the desired track shown on the CDI arrow-head. If the DI track indication changes you will start to move off track and the CDI indicator will move left or right showing you are right or left of track. The "bug" will start to move and by default indicates a course to steer to get rapidly back on track. The Author recommends reconfiguring the GPS to replace the default Vertical Speed to Target field with Bearing which tends to be much more useful. Use the Zoom in and out buttons to change the CDI scale sensitivity (the setting appears at the bottom left).

GPS III Pilot Highway Page

This unique screen gives a visual "highway in the sky" display. Its default settings are very good for getting you to the waypoint. If the road is straight ahead and you are on the centreline, you are on track! Use the Zoom in and out buttons to change the screen scale sensitivity (the setting appears at the bottom right).

GPSMAP 195 Map Page

The Map page shows the Desired Track (DTK) and Track "made good" (TRK). If you are on course the Track will be the same and equal the Desired Track on your flight log. If the Track does not match then you are heading off course. Once the errors become significant (and depending on the zoom setting) the plane will diverge from the line on the map. In Chapter 13 the Author recommended preferred settings for this screen, which include increasing the number of fields to eight. Amongst other things this enables you to configure one of the fields to be the Bearing to the next waypoint.

GPSMAP 195 HSI Page

The HSI page shows a lot of useful visual information as described in the panel above. If you are on course the CDI will be centred and the DI track indication will match the desired track shown on the CDI arrow head. If the DI track indication changes you will start to move off track and the CDI indicator will move left or right showing you are right or left of track. The "bug" will start to

Correcting Drift and Track Error — Detecting Track Error

move and by default indicates a course to steer to get rapidly back on track. By default one of the fields is configured to give the bearing to the next waypoint. As you move off course this will change to always give you a direct bearing to the waypoint. Use the Zoom in and out buttons to change the CDI scale sensitivity (the setting appears at the bottom right).

GPSMAP 295 Map Page
The Map page includes a copy of the HSI with much of the functionality described below. Chapter 10 recommended the Author's preferred settings for this screen and, because the HSI is readily available anyway, recommends setting this screen to have 8 conventional fields including, amongst other things, Course (desired track), Track ("made good") and bearing. With those settings if you are on course the Track will be the same and equal the Course (Desired Track) forecast on your flight log. If the Track does not match then you are heading off course. Once the errors become significant (and depending on the zoom setting) the plane will diverge from the line on the map.

GPSMAP 295 HSI Page
The HSI page shows a lot of useful visual information as described earlier. If you are on course the CDI will be centred and the DI track indication will match the desired track shown on the CDI arrow-head. If the DI track indication changes you will start to move off track and the CDI indicator will move left or right showing you are right or left of track. The "bug" will start to move and by default indicates a course to steer to get rapidly back on track. By default, the Track and Course are repeated in the digital fields and the track should match the course if you are to remain on track. The cross track error (XTK) is also shown. The Author recommends reconfiguring the GPS to replace the default Vertical Speed to Target (VST) field with Bearing which tends to be much more useful. Use the Zoom in and out buttons to change the CDI scale sensitivity (the setting appears at the bottom right of the HSI display).

Correcting Track Error
The first part of this chapter looked at the ways in which you might want to adjust your track in response to track errors. This part revisits these in the context of GPS. You will probably find these adjustments easiest to make using the HSI (or CDI) page as all the information will be at your finger tips. Alternatively you can use the Map page with default settings and your paper flight log, or better still, the Map page with the configuration changes that were recommended in Chapter 13.

The key information you need is Track (made good), Desired Track (course) and Bearing.

Adjust to parallel track

You probably don't really need to do this in the context of GPS, however once off track all you need to do is to turn so that the Track is the same as the Desired Track. If you are using the simulator you will need to make the turns on the HSI page (CDI page on GPS 92).

Re-acquire the route at the next waypoint

This is very straightforward - simply turn so that the Track and Bearing match. Remember that if you are using the simulator you will need to make the turns on the HSI page (CDI page on GPS 92).

> **Alert:** Remember that you are routing from an off track position directly to the waypoint, check your position on your paper chart and make sure you won't infringe airspace or compromise your MSA.

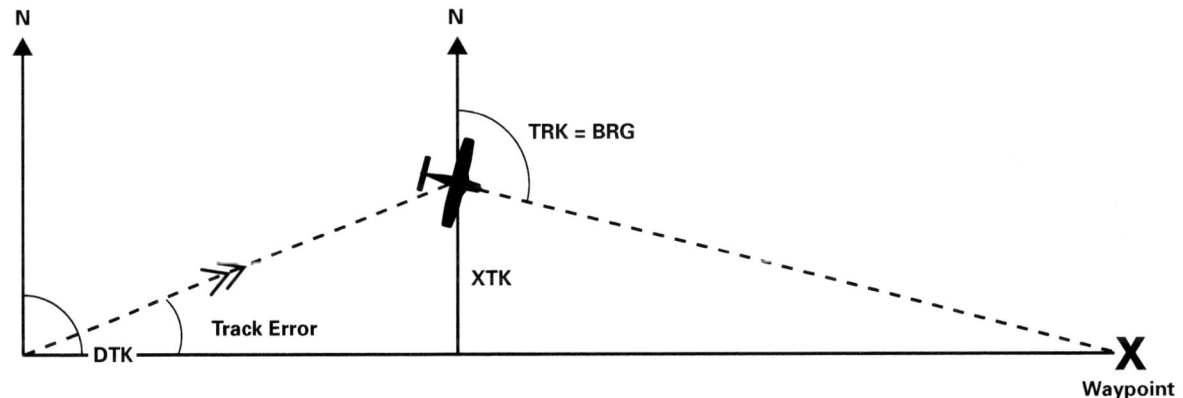

Re-acquire track before the next waypoint

There are two ways of doing this:

The first works on all models except the GPS 92 and is to simply follow the recommended course to steer by turning so that track matches the "bug" on the HSI screen. You can configure other GPS pages to display a course to steer field (CTS or To Course depending on model). The course to steer will increase the further off course you become to give a maximum intercept with the desired track of 45 degrees.

Correcting Drift and Track Error **Correcting Track Error**

> ⚠ **Alert:** The maximum course to steer generated by the GPS is 45 degrees. If you are a long way off track and/or quite close to the waypoint, it is possible that a 45 degree intercept will not regain track before the waypoint. You need to watch for this.

The second way is really identical to the traditional VFR method, except that the GPS provides you with so much information at your fingertips that you can exercise a greater degree of control over the process and choose the angle with which you wish to intercept the track.

First you need to choose an interception angle. Typically this will be similar to the angle which instrument pilot's use to intercept tracks to conventional radio aids. This will normally be between 30 degrees and 60 degrees, though if you are correcting only a minor cross track error you might want a much smaller angle, say 10 degrees.

What you do simply is to add or subtract the intercept angle to the desired track (depending if you are left or right of track) and turn so the course matches.

> 💬 **Tip:** Once you've turned, it's worth checking the track against the bearing to make sure your intercept will bring you back on track before the next waypoint!

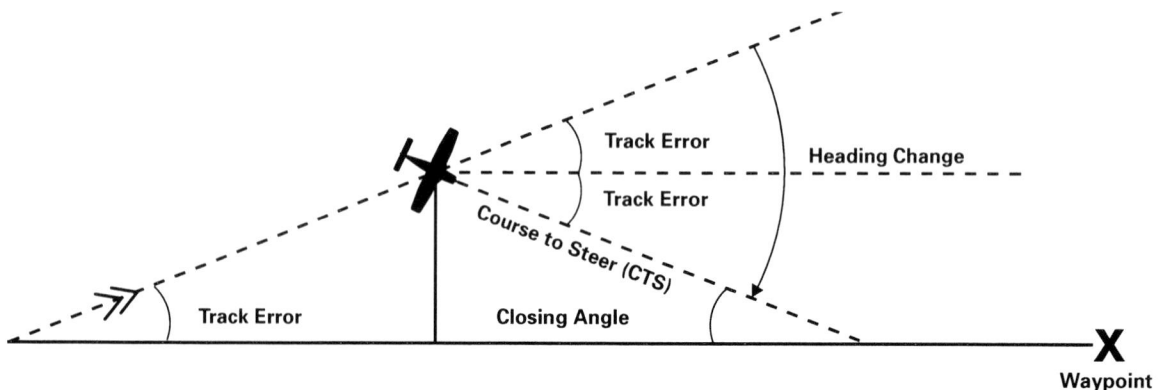

Illustration of re-acquiring track before the next Waypoint. The GPS will automatically choose a sensible closing angle and will display it in the CTS field (if you have one enabled) and by default the HSI bug will also show the course to steer. Note that the GPS will use a maximum intercept of 45 degrees - sometimes this might not be enough to re-acquire track before the waypoint, so be wary!

Correcting Drift and Track Error | **Correcting Track Error**

Maintaining Track using GPS

Earlier in this Chapter the possible reasons for getting off track were discussed. Without a GPS it can be difficult to be sure of the contributory effect of each of the factors: how well you hold your heading, the DI alignment and the actual wind aloft.

With GPS you can easily find out if the winds are as forecast:

Check the DI alignment (you might even consider using the compass correction card for the purposes of this exercise!) and concentrate on holding heading as precisely as possible. Monitor the Track as displayed on the GPS. If the winds are as forecast then the Track (made good) should match the Desired Track/Course (which should match what's on your flight log!). If it does not then the track will vary and over time you would drift off course.

By checking the DI and heading holding you therefore can modify your planned heading to ensure there is no drift - turn until Track (made good) and Desired Track/Course match again - now check the heading indicated on the plane's DI (NOT the GPS CDI/HSI or compass display). If you now steer this heading on this leg you should stay on track!

Of course if the wind direction has changed, the forecast ground speed will also probably have changed. You can compare the actual ground speed from the GPS with the forecast speed on your flight log to see how great the difference is, and what kind of adjustment to the ETA to expect at the next waypoint. Of course the route page of the GPS will be re-forecasting the ETA for you all the time based on live information!

Winds Aloft

Using the GPS you can actually find out what the winds are. This is normally more of interest than of navigational use since you can easily find the required wind correction angle and ground speed as described above.

Airmanship: Using this function tends to cause a "lot of head in cockpit". Be sure to exercise appropriate lookout between each step, or preferably use a safety pilot.

Computing the Winds Aloft

GPS 92	GPS III Pilot	GPSMAP 195	GPSMAP 295
From the Main Menu select **E6-B Menu** and **Winds Aloft.** If your ASI has a TAS adjustment, compute it using the actual outside air temperature (alternatively you can use the Density Altitude (and TAS) menu from the E6-B menu to compute TAS). Enter your current heading (as shown on the aircraft's DI correctly aligned and corrected for compass deviation). You MUST steer the heading you enter! The wind speed and direction will be computed as well as the headwind component.	From the Main Menu select **E6B Menu** and **Winds** tab. If your ASI has a TAS adjustment, compute it using the actual outside air temperature (alternatively you can use the Density Altitude (and TAS) tab to compute TAS). Enter your current heading (as shown on the aircraft's DI correctly aligned and corrected for compass deviation). You MUST steer the heading you enter! The wind speed and direction will be computed as well as the headwind component.	From the Main Menu select **Density Alt/Winds Aloft.** If your ASI has a TAS adjustment, compute it using the actual outside air temperature (alternatively you can use the Density Altitude and TAS computation in the top part of the screen). Enter your current heading (as shown on the aircraft's DI correctly aligned and corrected for compass deviation). You MUST steer the heading you enter! The wind speed and direction will be computed as well as the headwind component.	From the Main Menu select the **E6B** tab. If your ASI has a TAS adjustment, compute it using the actual outside air temperature (alternatively you can use the Density Altitude and TAS computation in the top part of the screen). Enter your current heading (as shown on the aircraft's DI correctly aligned and corrected for compass deviation). You MUST steer the heading you enter! The wind speed and direction will be computed as well as the headwind component.

Introduction

In this Chapter you will learn how the GPS can help you if you need to divert. The Chapter also considers what to do if you have been vectored off your planned route by ATC and then given the all clear to "resume own navigation". Finally you will learn how to reverse (invert) your route and will also be introduced to the use of "TracBack".

Diverting to the Nearest Airfield

There are numerous reasons why you may wish to divert - for example, in the event of worsening weather, an airsick passenger or a rough running engine. A GPS will help you considerably in the event you need to make such a diversion, firstly by locating the nearest suitable airfield and secondly by planning the route.

The steps you will follow are very much the same as you would follow using conventional navigation however almost all the hard work can be eased using the GPS:

- Confirm the current location
- Identify the new destination
- Draw a line on the aeronautical chart
- Measure the distance and course
- Compute the required heading to make good the new track
- Compute the speed and ETA to the destination

Confirm the Current Location

Using the GPS this should be very straightforward. Assuming you are on course, you should be on a clearly defined point on your track, and, unless you have configured it otherwise, your GPS will be displaying the distance to the next planned waypoint. Using this information you quickly can measure and identify your exact position on the chart.

Identify the New Destination

The GPS makes this very easy for you. You can use the "nearest" function:

Diverting to Nearest Airfield			
GPS 92	**GPS III Pilot**	**GPSMAP 195**	**GPSMAP 295**
Press the [Goto/NRST] button twice to activate the "nearest" function.	Press and hold down the [Goto/NRST] button to activate the "nearest" function.	Press the [NRST] button to activate the "nearest" function.	Press the [NRST] button to activate the "nearest" function.
A list of the nearest airports, their distances and bearings should be displayed. If another data type happens to be displayed, press the [Goto/NRST] button again to select airports (**APT**).	A list of the nearest airports, their distances and bearings should be displayed. If another data type happens to be displayed, press and hold [Goto/NRST] again to select airports.	A list of the nearest airports, their distances, bearings and runway lengths should be displayed. If another data type happens to be displayed, press the [NRST] button again to select airports.	A list of the nearest airports, their distances, bearings, runway lengths and elevation should be displayed. If another data type happens to be displayed, press the [NRST] button again to select airports.
Select a suitable airfield using the rocker pad up/down and press [Enter].	Select a suitable airfield using the rocker pad up/down and press [Enter].	Select a suitable airfield using the rocker pad up/down and press [Enter].	Select a suitable airfield using the rocker pad up/down and press [Enter].
This will enable you to view the airfield data including frequencies and runway data. You can press [Quit] to select an alternative if you are unhappy with your selection.	This will enable you to view the airfield data including frequencies and runway data. You can press [Quit] to select an alternative if you are unhappy with your selection.	This will enable you to view the airfield data including frequencies and runway data. You can press [Quit] to select an alternative if you are unhappy with your selection.	This will enable you to view the airfield data including frequencies and runway data.
Copy the frequencies and runway data to your paper flight log, just in case you can't find a plate in your flight guide.	Copy the frequencies and runway data to your paper flight log, just in case you can't find a plate in your flight guide.	Copy the frequencies and runway data to your paper flight log, just in case you can't find a plate in your flight guide.	You can press [Quit] to select an alternative if you are unhappy with your selection.
When you are happy, press [Goto] again and then [Enter].	When you are happy, press [Goto] again and then [Enter].	When you are happy, press [Goto] and then [Enter].	Copy the frequencies and runway data to your paper flight log, just in case you can't find a plate in your flight guide.
			When you are happy, press [Goto] and then [Enter].

Nearest Airport Page

NEAREST AIRPORTS			
WAYPOINT	BRG	DIS	LENGTH
EGLD	223°	1.6%	2500'
EGWU	146°	4.2%	5500'
EGTR	068°	6.6%	2100'
EGLL	177°	7.9%	12800'
EGTB	274°	12%	2400'
EGLM	242°	13%	3600'
EGWN	323°	14%	3700'
EGTF	193°	16%	2600'
EGGW	019°	17%	7000'

GPS 92   **GPS III Pilot**   **GPSMAP 195**   **GPSMAP 295**

Tip: If you are heading into deteriorating weather you can use the bearing information on the nearest page to choose an airfield that is behind you or at least away from the weather.

Fact: You can configure the GPS to limit the display of airfields to those of specific runway types and lengths. By default most light aircraft can get into any airfield in the database in an emergency so at this stage of this book configuring the minimum runway length is left as an exercise for the reader.

Draw a Line on the Chart

Reconfirm your exact position (if the step above took you a minute or so you may have travelled a couple of miles). Draw a line on the chart from your current position to the new destination.

Measure the Distance and Course

The GPS will have done this automatically for you when you executed the Goto. You can transfer the Distance and Desired Track/Course directly from the Active Goto page.

Compute the Required Heading to Make Good the New Track

At this stage you would normally have to do quite a lot of measurement and estimation. The GPS enables you to sidestep this:

Diversions using GPS   **Diverting to the Nearest Airfield**

Turn towards the waypoint and start to track towards it ensuring Track (made good) equals Desired Track/Course equals Bearing. Provided you execute the turn safely but expediently they should not diverge excessively.

Using what you learned in the previous Chapter you will be able to soon be able to track accurately and establish the wind correction needed: as soon as you have established the track and it's stable, check the aircraft's DI alignment and heading so you know what heading is required to maintain the track in the prevailing wind conditions. Note this heading onto the flight log (just in case the GPS should fail).

Compute the speed and ETA to the destination

Note the Ground speed from the GPS and the ETA for the destination (or compute the ETA from the current time and the ETE if more convenient).

> **Airmanship:** When you have finished do a "gross error check" - did you select the right waypoint? Is the drift and groundspeed consistent with the wind? Does the ETA make sense?

Diverting Around Weather

The GPS can be used to assist in diversions around weather. Most instructors and training publications teach either or both of the 45 degree rule and the 60 degree rule: divert around the weather (for example a heavy rain shower) in your path by heading away from track towards a clearer area by 45 or 60 degrees.

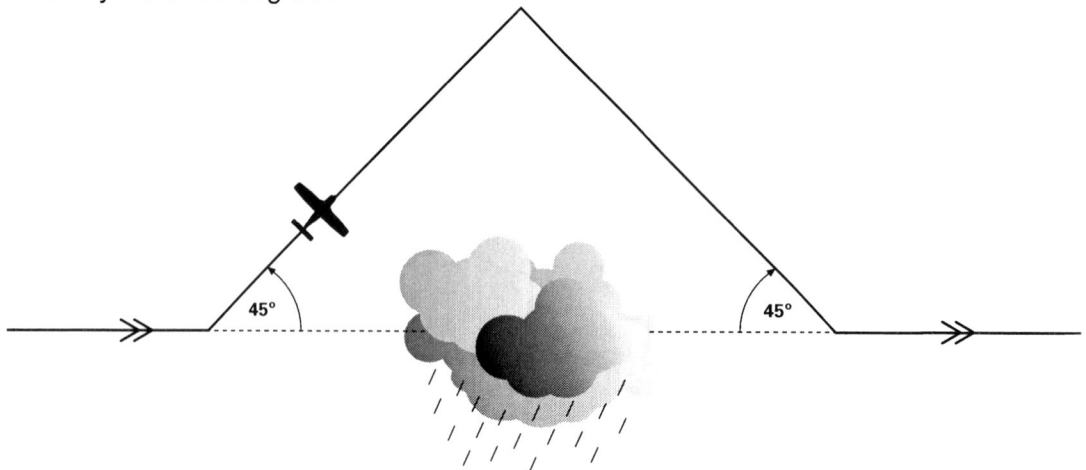

Diverting around a shower using the 45° method.

The 45 degree method has the advantage that it is easier to do "visually". The 60 degree method has the advantage that it means that the two sides form the sides of an equilateral triangle and therefore the diversion is twice the length and time (in still air) as the segment that is diverted around, making re-computing estimates easier.

Either method can have a parallel segment inserted if required.

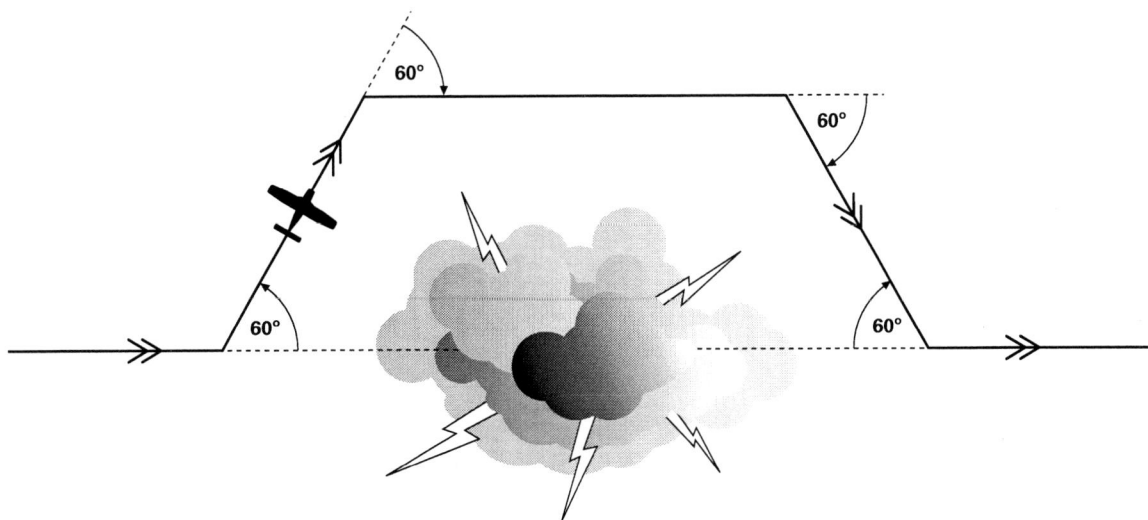

Diverting around a thunderstorm using the 60° method coupled with a parallel segment.

Using the GPS you can very easily fly these diversions accurately as well as having an alternative option of heading straight to the next waypoint when you are clear of the weather:

Simply fly off track so that the track ("made good") is 45 degrees or 60 degrees as required from the desired track/course. When the weather has been passed, turn back in order to achieve a 45 or 60 degree intercept as explained in the previous Chapter. Alternatively using the GPS it is very easy to follow the bearing direct to the waypoint or choose another intercept angle if required.

If a parallel segment is required, simply match the track to desired track/course.

Alert: Ensure that your revised track won't take you through any airspace or compromise your MSA!

Alert: If the weather is too bad you may need to turn back using the method described below, or diverting to another airfield as described above.

Diverting Around Danger Areas

Ideally you will have planned your route to avoid danger areas, however if you were planning to get a clearance through them and find that one is not available, or you have intended to over-fly them and the cloud-base has prevented you, then you may need to divert around them.

In general the way to do this is exactly the same as diverting around weather using the 45 degree or 60 degree method as described above.

Tip: It is very tempting to use the GPS to "skim" around danger areas. This is reasonably valid for very small circular areas in conjunction with visual checks against the ground and chart, however Air Traffic Services tend to get "nervous" if you do this with larger areas. Plan to be a reasonable distance from larger danger areas using a parallel segment in the diversion if required.

Alert: The Jeppesen database does NOT include High Intensity Radio Transmission Areas (HIRTAs).

"Resume Own Navigation"

Before very long in your flying career you may find yourself vectored around by a controller and then instructed to "resume own navigation". You are perfectly within your rights, of course, to ask for further information such as a vector to your next waypoint. With the GPS, that becomes unnecessary because you can instantly know the distance and bearing. Using the techniques discussed in the last Chapter you can easily find your way.

Alert: Ensure your revised track won't take you through any airspace or compromise your MSA!

Diverting Back along your Track

Sometimes you may wish to abandon a flight and return to your departure point - in fact these kinds of diversion are probably more common than diverting to an unknown airfield - after all, who wants to be stuck 100 miles from home unless absolutely necessary?

Alternatively you might have seen something on the ground (maybe an incident, or just a point of interest) and wish to return to it to report it or photograph it.

There are two main ways of doing this:

- Reversing your route: If you have remained on track, you can "invert" your route and follow it backwards, waypoint by waypoint.

- "TracBack": By default the Garmin GPS stores a trail of points representing your actual track over the ground. You can automatically create a route to follow that track.

Reversing (Inverting) your Route

You can very quickly reverse the route and then all you have to do is follow it.

Reversing (Inverting) the Route			
GPS 92	**GPS III Pilot**	**GPSMAP 195**	**GPSMAP 295**
On the Active Route page, select Invert (**INV**) and press [Enter]. The GPS is smart enough to know which leg you are closest to and should activate it as the current leg.	On the Active Route page, press [Menu] to activate the context sensitive menu, select **Invert** and press [Enter]. The GPS is smart enough to know which leg you are closest to and should activate it as the current leg.	On the Active Route page, press [Menu] to activate the context sensitive menu, select **Invert Route** and press [Enter]. The GPS is smart enough to know which leg you are closest to and should activate it as the current leg.	On the Active Route page, press [Menu] to activate the context sensitive menu, select **Invert** and press [Enter]. The GPS is smart enough to know which leg you are closest to and should activate it as the current leg.

Fact: You can also invert routes on the Route planning pages and of course you can invert a route on the ground at your destination if you want to take the same route home in reverse.

Using "TracBack"

The GPS stores a track of your actual track over the ground. This is normally shown on the map page as a dotted line (solid line on the GPS 92). To configure the Track Log and use "TracBack" proceed as follows:

Using "TracBack"			
GPS 92	**GPS III Pilot**	**GPSMAP 195**	**GPSMAP 295**
From the Map page, select the options menu **(OPT)** and select **Track Log**. To activate TracBack, select **TracBack** and press [Enter]. The GPS will create a series of waypoints representing the track prefixed by T and create a new active route using them. This process can take a minute or so.	From the Main Menu select **Track Log**. To activate TracBack, select **TracBack** and press [Enter]. The GPS will create a series of waypoints representing the track prefixed by T and create a new active route using them. This process can take a minute or so.	From the Main Menu select **Track**. To activate Tracback, select **Start TracBack** and press [Enter]. The GPS will create a series of waypoints representing the track called SYMBOL and create a new active route using them. This process can take a minute or so.	From the Main Menu select the **Track Log** tab. To activate Tracback, select **TracBack** and press [Enter]. The GPS will create a series of waypoints representing the track called TURN and create a new active route using them. This process can take a minute or so.

Configuring the Track Log

From the applicable Track Log menu you can **Clear** the Log and set up its resolution and whether it is **Off** or will **Wrap** (overwrite early parts of the log when the track log gets full) or **Fill** (in which case recording will stop with a warning message when the Log is full).

You can also vary the resolution of the log - the default settings are normally OK unless you want to record a very long trip.

Tip: Using Tracback can result in a large number of later unwanted user waypoints on some of the GPS models. You can delete the TracBack waypoints from the context sensitive menu on the applicable user waypoint page.

4

GPS as an aid to Radio Navigation

This section is aimed at more experienced pilots, especially those already skilled in Radio Navigation such as IMC rated and IR pilots.

It is worth reinforcing the point that Garmin portable receivers are only intended as a supplementary navigation tool for VFR flight, however they can provide a valuable source of situational awareness for pilots engaged in Radio Navigation.

This section anticipates the use of the GPS III Pilot, GPSMAP 195 or GPSMAP 295. The GPS 92 can still be used to assist with en route navigation but lacks some of the more advanced features discussed in this section.

Introduction

In this Chapter you will continue the trip northwards from Manchester Barton to Edinburgh.
The first part of this Chapter reinforces what you learned in Section 3, however some new ideas
are introduced including, for GPSMAP 295 owners, vectored approaches to a major airport.

You will fly the first part of this trip up the coast past Blackpool and then over the Lake District
direct to Edinburgh. Since you are qualified to use radio navigation you will use the convenient
NDB's and VORs.

You will need to use the Northern England and Northern Ireland 1:500000 chart.

> ⚠ **Alert:** If you find yourself needing to constantly refer back to Section 3 to
> remind yourself how to carry out the various functions discussed here you
> probably aren't ready to take your GPS into the air yet.

Planning the Route

1. Choose the Waypoints.

You will route from Manchester Barton (EGCB) to
Blackpool NDB (BPL), to Dean Cross VOR (DCS), to
Talla VOR (TLA), and then to Edinburgh (EGPH).

2. Minimum Safe Altitude

Now compute the minimum safe altitudes. You have
some fairly inhospitable territory and some high MSAs.

FROM / TO	MSA	PL/ALT	TAS	TR (T)	W/V	HDG (T)	HDG (M)	G/S	DIST	TIME	ETA	ATA
EGCB / BPL	3500				/							
BPL / DCS	4300				/							
DCS / TLA	3800				/							
TLA / EGPH	3800				/							

3. Plan your Altitude

Plan to climb to Flight Level 45 after Blackpool and then to Flight Level 50 to give you clearance over the higher ground and to keep the correct quadrantal levels for the directions you are heading.

4. Entering the Route

Now you are ready to enter the route into the GPS:

Enter the route:

- EGCB
- BPL
- DCS
- TLA
- EGPH

Fact: If you need a recap on entering a route, refer back to Chapter 13.

5. Measure the Leg Distances

Now you can enter the leg distances in your flight log. Simply look down the Route Page on your GPS and enter in the leg distance figures. Enter the distance on your flight log. Except for the GPS 92, the GPS displays a decimal figure: round to the nearest nautical mile.

Alert: If you've got a GPS III Pilot you may need to use the left/right rocker function to scroll the displayed field. Be sure to use the **Leg Dist** display and not the **Distance** display (which is the cumulative distance at each point).

Now you ought to do a "gross error check". Put a ruler to each line on the chart to check quickly that the distances are correct. The GPS means that you don't need to take time measuring the lines accurately with a ruler, but it's still worth doing a quick check.

6-8. True Track, True Airspeed, True Heading and Magnetic Heading

Next you can compute the wind effect on your speed and work out the heading required to "make good" the track for each leg. You should compute all the legs in full as you did in Section 3.

> **Tip:** It's useful to have a flight log that shows Magnetic Heading and Magnetic Track. It's worth considering compiling your own preferred layout of a flight log using a word-processor or spread-sheet if you can't get what you want on a commercially available flight log.

9 & 10. Compute your Time and Fuel requirement

Compute the fuel requirement for each leg based on the stage of flight and the published fuel settings for your aircraft.

Remember to add Taxi fuel (normally a couple of gallons/10 litres) and holding/diversion time (normally 30-45 minutes) to get the full requirement.

> **Alert:** Experienced pilots are often tempted to skip computing all the legs and just estimate the time and fuel requirement. Good practice dictates that you should compute the flight log in detail just as for a normal VFR flight not using radio aids.

> **Tip:** If you are going to ignore the advice above, you might be persuaded to use the total distances and "average" desired track /course from the GPS route plan, do the wind computation for this "virtual" leg and compute a better time and fuel estimate. This can be quite accurate for cross-country routes in a single general direction.

11. Fan Lines ("track guides")

You are probably experienced enough to do without these.

12. Track Markings

Add 6 minute markers, mileage markers or fractional markers as you prefer.

DATE / /		FUEL CONSUMPTION		
PILOT	AIRCRAFT	TOTAL REQUIRED		
FROM	TO	FUEL ON BOARD		
DISTANCE	FLIGHT TIME	RESERVE		
START UP	TAKE OFF	TOTAL ENDURANCE		
LANDING	SHUT DOWN	COMMUNICATIONS & RADIO NAV. INFORMATION		
ALTERNATE		DISTRESS 121.50		
DISTANCE	FLIGHT TIME	STATION	SERVICE	FREQUENCY
SUNSET	VARIATION			
2000' W/V 150 / 15	TEMP. +10			
5000' W/V 150 / 15	TEMP. +4			
DEPARTURE INFORMATION				

FROM / TO	MSA	PL/ALT	TAS	TR (T)	W/V	HDG (T)	HDG (M)	G/S	DIST	TIME	ETA	ATA
EGCB / BPL	3500	FL 45	106	308	/	304	309	120	29	15		
BPL / DCS	4300	FL 45	106	348	/	350	355	126	58	28		
DCS / TLA	3800	FL 50	107	359	/	002	007	127	47	22		
TLA / EGPH	3800	FL 50	107	358	/	001	006	127	27	13		

Finishing off...

Now you are ready to go. Complete your flight log with the frequencies of the navaids and ground stations you will talk to. You'll need PPR for Edinburgh. You may also want to file a flight plan for this route as it is over fairly inhospitable territory. Consider an alternate. Don't forget to activate the route on the GPS.

Basics of GPS and Radio Navigation   **Planning the Route**

Flying the Route on the Simulator

Enable Simulator mode and initialise the position to Manchester Barton (EGCB).

Set the altitude (from the Satellite Status page menu) and on the HSI page accelerate the Simulator (taking care not to steer it manually at the same time). You may want to follow along on an aeronautical chart and you may wish to accelerate to super fast Simulator speeds for the first part of the route!

Consider setting a VNav profile (in real life you would do this before departure).

As you get past Talla (TLA) slow down to a more conventional 100Kts.

You will soon be talking to Edinburgh Approach and can get the ATIS. You should continue to track towards Edinburgh.

The Jeppesen database in the GPSMAP 195 and GPSMAP 295 features the final approach segment for airfields with instrument approaches. The final approach segment normally starts at the Final Approach Fix (FAF) and continues to the Runway or Missed Approach Point (MAP) depending on the approach type.

> **Fact:** Unlike panel mount Garmin GPS units, the portable units do NOT contain the full approaches.

> **Note:** To confuse things further, some approaches in fact don't include the entire final approach segment. The Edinburgh approaches to Runway 24 actually have their FAF at 6.1nm, whilst the GPS only shows the portion from the EDN beacon inbound.

Approaches are considered further in Chapter 20, however the GPSMAP 295 has a special feature to help with situational awareness if you are being vectored to the approach by ATC:

Vectored Approaches (GPSMAP 295 only)

Although you have been using radio navigation, this is officially a VFR flight so it is likely that you will receive vectors to the approach as you get closer. Once you are being vectored you might want to activate an approach to the runway in use.

Go to the Active Route page and press (MENU) once to activate the context sensitive menu and choose **Select Approach**. This will display the available approaches for the airfield at the end of the route, in this case Edinburgh.

In this example, assume that you are being vectored to a landing on runway 24, so select **ILS RW24.**

The GPS will ask if you are receiving **Vectors**? If you answer **Yes**, it is assumed that you are under radar vectoring and so you will no longer receive GPS routing information from your current position. The GPS creates an active route line from the FAF to the runway threshold (or MAP).

If you had answered **No** to the question about vectors, the destination waypoint will be substituted by waypoints representing the FAF and the runway threshold (or MAP) and you will be routed initially to the FAF. This is discussed in Chapter 20.

Fact: If the weather is good and you have the airfield in sight you might consider not bothering with selecting an approach on the GPS and looking out of the window instead! Remember also that the FAF is designed for an instrument approach and for a visual approach ATC may well vector you somewhere midway along the final approach segment.

Tip: If you are on a VFR flight, it is best to choose an ILS approach rather than an NDB approach as the final approach segment generally goes to the threshold rather than a missed approach point.

Alert: Don't activate vectoring until you are actually being vectored as the active route line is set to the final approach segment, which may still be some way away. If you inadvertently activate the approach too early, select **Remove Approach** from the Active Route page menu.

Simulator: As soon as you select vectors in Simulator mode you will normally have to start to "fly" the aircraft manually, since the Simulator isn't clever enough to intercept the final approach segment and will probably fly off at a 45 degree tangent, trying to intercept the extended track.

Flying the Route for Real

You can now try flying a route for real. Here are the things you should look out for again:

Alert: The first time you use the GPS for real go up with a safety pilot, or better still sit in the right hand seat and do the navigating before progressing to the left hand seat.

Checklist: It's a good idea to switch on your GPS at the same time you switch on your other avionics. This is normally shortly after you have started the engine. Remember to check the database validity and allow sufficient time for the receiver to acquire satellites.

Bracket the track easily using GPS: Fly using the navaids with a suitable wind adjustment to achieve the desired track. Don't "fly" the line on the GPS. You can use the heading adjusted for wind that you (hopefully) computed during the planning stage as a starting point and use the GPS track display to match the desired track/course to work out the actual drift.

Checklist: When you do your cruise checks is a good time to check the GPS is functioning properly. Page to the Satellite Status screen. Do you have good coverage? Is the Battery OK (or are you still on external power)? Does the waypoint the GPS is aiming at match the one you are really aiming at?

Tip: When you are airborne things are very different from on the ground and you may find you temporarily forget some of your GPS training! The following tip works anywhere to get you back to the Satellite Status page: Just keep on pressing QUIT . No matter where you are, you will soon be back to the Satellite Status page.

Alert: Once you have the destination airfield in sight, STOP fiddling around with the GPS and make sure you have done your airfield approach checks and in due course your pre-landing checks.

Checklist: As you shut the aircraft down you will probably switch the avionics off. Now is a good time to switch off the GPS.

Introduction

This Chapter looks at a useful, but little used feature of Garmin portable GPS receivers: OBS mode. OBS stands for Omni Bearing Selector and offers similar functionality to adjusting the OBS knob on a real aircraft CDI, tuned into a VOR.

OBS mode is not available on the GPS 92.

There are relatively few instances in which you are likely to need to use OBS mode in real life, however it is introduced in this Chapter partly for completeness and partly because many Garmin panel mount GPS receivers you might also encounter use OBS mode more extensively, so a good understanding on the portable may bring later benefits.

In short, the OBS function enables you to visualise a track following a radial inbound to a VOR, NDB (or other waypoint).

Tip for GPSMAP 295 users: Using OBS mode is one of the few instances in which the Author would consider setting the right hand area of the Map page to include the HSI display.

NDB Tracking using OBS Mode

The most common use of OBS mode is to assist with intercepting a radial inbound to an NDB, where that radial is not on your current track. The GPS can assist you in visualising the interception.

Fact: If you are not trained in radio navigation or IMC rated you may be puzzled by the comment that NDBs can have radials. A trained pilot can interpret the position of the ADF needle, the aircraft's heading and take account of the wind effect to build a mental picture that lets him or her fly to an NDB as if it had radials like a VOR. Explaining how this is done is deliberately beyond the scope of this book, however using a GPS can be of significant help to trained pilots in building this mental picture.

Imagine you are westbound out of Elstree and want to achieve a 45 degree intercept of the 320 radial inbound to Westcott (WCO):

Switch Simulator mode on and initialise the position to Elstree (EGTR). Set a speed of say 100kts.

On the GPS activate a Goto WCO:

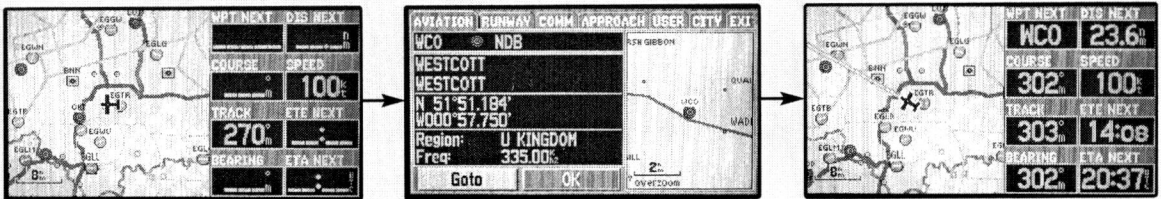

GOTO WCO

This creates a track line from your current position - this is not what you want, so you have to use OBS mode...

On the HSI page, press (MENU) to activate the context sensitive menu and select Set **OBS and Hold**. The track/radial will appear on the HSI display. Use the rocker pad up/down to change the radial until it reads 320° and press (ENTER).

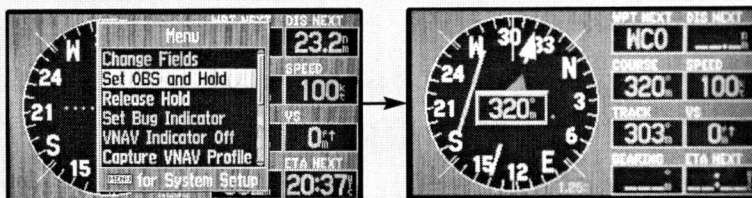

Setting up OBS Mode

Intercepting the Radial

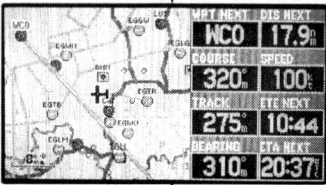

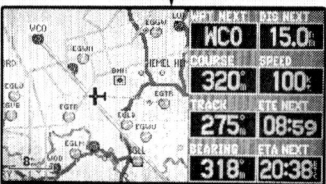

You would now fly the plane tracking 275° to give you the 45° intercept (320-45=275).

You will notice the CDI D-bar move out to the left of track. If you (QUIT) to the Map page you will notice that the active track line is now to the west of you and follows a radial of 320° inbound to the NDB.

In the real aircraft you will also have tuned in Westcott on the ADF and identified it.

Fly a steady heading of 275° (correcting for wind if required in real life) until your ADF needle (or RMI) is showing that you have achieved the 45° interception. At this point you will find you have reached the track line on the map and the GPS HSI D-bar has become centred.

In fact what you have done is turned the NDB into a VOR! You should of course continue to fly using the aircraft's ADF (or RMI) as your primary navigation, however you can use the map display and the GPS HSI for additional situational awareness.

As you track inbound to the NDB, allowing for wind and maintaining ("bracketing") track is made very much easier because the GPS displays your Track "made good" at all times and this should of course be 320°.

You can try flying this on the Simulator!

To stop using OBS mode, press (MENU) on the HSI page to activate the context sensitive menu and select **Release Hold**.

Visualising Holds

OBS mode can be used to visualise holds, using a very similar method to the NDB tracking technique described above. Explaining how to fly a hold in real life is deliberately beyond the scope of this book.

With the NDB (or other hold fix) as the next active waypoint (either as part of a route or as a Goto), choose the HSI page, press (MENU) to activate the context sensitive menu and select **Set OBS and Hold**. The track/radial will appear on the HSI display. Use the rocker pad up/down to change the radial until it matches the inbound radial for the hold and press (ENTER).

You can now use normal techniques to fly the hold, however as you turn back inbound, you now have a course line and HSI to help you visualise the interception and tracking of the inbound course.

> **Alert:** You should not use the GPS to fly an exact racetrack pattern in windy conditions - you should still fly the hold exactly as you would if the GPS wasn't present.

> **Tip:** When learning to do holds and appreciate the effect of wind it can be particularly instructive to ask your instructor to operate the GPS (out of sight of yourself) so it records your track which you can review later on the ground and see how the racetrack pattern is affected.

0

Using OBS as an alternative to adding a user defined waypoint

Imagine that you are flying from Southampton to Kemble. The direct route will take you across Boscombe Down and danger areas that are frequently active to high altitudes.

Instead you may wish to travel via Membury Mast, which, together with the adjacent M4 service station, makes a reasonable visual waypoint.

Rather than create a user defined waypoint as described in Chapter 15 and enter it into a route, you may wish to simply fly to the waypoint visually with the GPS providing additional situational awareness.

To do this, all you have to do is plan your two leg route on a flight log as usual. On the GPS create a "Goto" SAM. This may seem strange as you are already at SAM (which is at the south east corner of Southampton airport), however when you enter OBS mode a course line on the map will be created, which extends both North and South of SAM.

Next, set up the OBS course: On the HSI page, press (MENU) to activate the context sensitive menu and select **Set OBS and Hold**. The track/radial will appear on the HSI display. Use the rocker pad up/down to change the radial until it reads 350° and press (ENTER).

As you travel away from Southampton you will see that the direction arrow on the HSI display points backwards towards SAM - this is one of the only times it does this.

If you are used to using radio navigation aids, you will probably set up SAM on a NAV radio and set 350 degrees on the aircraft's CDI (or HSI) too.

When you reach Membury you need to press (MENU) on the HSI page to activate the context sensitive menu and select **Release Hold**. Then you can do a straightforward Goto to Kemble (EGBP).

Using OBS to Avoid Airspace

Imagine that you are starting a flight from Denham and your first leg would ideally route towards Compton VOR (CPT). Denham lies just within the Northern edge of the London CTR.

You could go to the trouble of getting a Special VFR clearance provided the weather limits are good enough, however this is a lot of trouble just to avoid snicking the north eastern part of the zone.

Direct Track to CPT

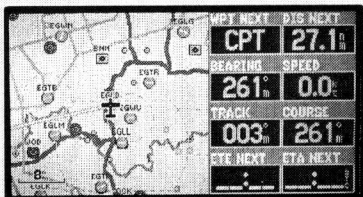

The direct track from Denham to CPT infringes Heathrow's airspace.

You could fly up to Bovingdon (BNN) first to keep you clear, however this would take you an unnecessary distance north and require quite a large turn when you got there.

In practice a pilot will most likely look to intercept a suitable radial inbound to CPT after departure from Denham. The 250° radial will do the job.

You can't easily cater for this on the GPS: if you use Goto then you would snick the CTR. You could enter a route (e.g. EGLD to CHT to EGTB to CPT), but this is tedious. However you can solve the problem using OBS.

To use OBS for this problem, assume you have planned a route with the first en-route waypoint being CPT. Switch Simulator mode on and initialise the position to Denham (EGLD). Enter a route into the GPS (EGLD to CPT to some other nearby point), or just do a Goto CPT.

Now, on the HSI page, press **MENU** to activate the context sensitive menu and select **Set OBS and Hold**. The track/radial will appear on the HSI display. Use the rocker pad up/down to change the radial until it reads 250°and press **ENTER** .

Setting the 250° Radial

The 250° Radial will keep you clear of infringing the London CTR.

You will notice the CD-I bar move out to the right of track. If you (QUIT) to the map page you will notice that the active track line is now to the north of you and follows the 250 radial inbound to CPT missing the London CTR and taking you neatly overhead Wycombe (EGTB).

Map and HSI Displays after Enabling OBS Mode

Upon preparing for departure from Denham you will set up the real aircraft NAV 1 to 114.35 (CPT) and adjust the VOR CDI compass card to 250 using its OBS knob.

After clearing the London zone you can turn left heading say 280° to achieve a 30 degree intercept to the 250 radial. IMC pilots are used to building this image in their heads. Using OBS mode it is displayed visually on the GPS Map as well.

You can fly this route on the Simulator (though the Simulator will do an overzealous 45 degree intercept unless you take over and fly manually).

Intercepting the Radial

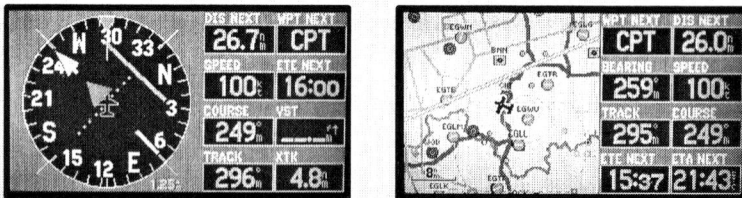

When you get to CPT you will want to continue on your route and have the waypoints automatically sequence again. To do this, press (MENU) on the HSI page to activate the context sensitive menu and select **Release Hold**. You must remember to do this otherwise you will continue to be guided outbound on the same radial.

OBS Mode   **OBS to Avoid Airspace**

Introduction

This book wouldn't be complete without including a Chapter on instrument approaches. A notion of instrument approaches is featured in the GPSMAP 195 and GPSMAP 295. For reasons that will become apparent, this Chapter is presented as a discussion. The Author leaves it up to more experienced pilots to make up their own mind as to whether or not to use this feature.

Discussion

Readers should consider very carefully whether or not to use the approach facilities in practice for a number of reasons:

- Only the final part of the approach is included on the Portable units.

- The database is inconsistent. Some approaches contain the full final approach segment from the Final Approach Fix (FAF) to threshold or Missed Approach Point (MAP) as applicable. Others only include the last part of the final approach segment.

- Most approaches in the UK start from an Initial Approach Fix (IAF). In practice you are very unlikely to either choose to or be allowed by ATC to self-position to the FAF.

- Garmin's own advice and the Jeppesen database licence provide that the GPS is for supplementary use in VFR conditions only.

- Many low hour IMC pilots may find that configuring the GPS for additional situational awareness during approaches adds more workload than it alleviates. They may be better off just interpreting the GPS Map page "visually" as they fly the approach using the appropriate radio navigation aids.

Having said all this, the remainder of this Chapter illustrates how to use the GPSMAP 195 and GPSMAP 295 to improve situational awareness during instrument approaches, and explains some of the assertions above in more detail. You learned about vectored approaches in Chapter 18, so this Chapter looks at procedural approaches.

The approaches into Edinburgh from the last route are a little complex and, as already mentioned, you will almost definitely be vectored onto the approach anyway. Consider some training approaches into Southend instead:

NDB Approach (Southend RW06)

Enable Simulation mode and set up a route (or a Goto) to Southend (EGMC). Go to the Active Route page and press **MENU** once to activate the context sensitive menu and choose **Select Approach**. This will display the available approaches for Southend. Select **NDB RW06**. If you have a GPSMAP 295 answer **No** to the question about Vectors.

EGMC will be replaced by the final approach fix and the missed approach point.

When was the last time you commenced an NDB approach directly from the Final Approach Fix? Probably never. You're not going to get much benefit from using the GPS like this.

ILS Approach (Southend RW24)

Enable Simulation mode and set up a route (or a Goto) to Southend (EGMC). Go to the Active Route page and press (MENU) once to activate the context sensitive menu and choose Select Approach. This will display the available approaches for Southend. Select **ILS RW24**. If you have a GPSMAP 295 answer **No** to the question about Vectors.

Selecting the Approach

EGMC will be replaced by the final approach point and the runway threshold.

Map Views after Selecting the Approach

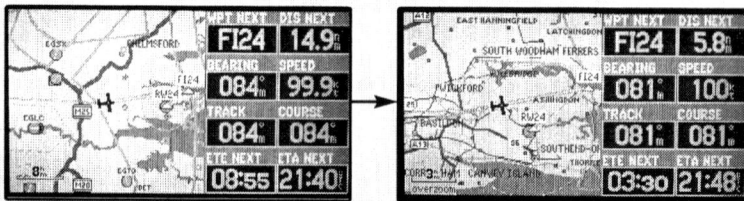

When was the last time you commenced an ILS approach from the Final Approach Fix? Probably never, as you would have carried out a procedural approach or been vectored onto the ILS.

Comment: In the Author's view the approaches stored in the GPSMAP models are not very useful in the UK where you will either carry out a procedural approach or be vectored onto an ILS. However the next part of this Chapter shows a trick you can use to improve things for procedural approaches.

Pre-preparing the GPS for Procedural Approaches.

In order to get some additional situational information from the GPS for use in procedural approaches you need to prepare further in advance, preferably on the ground before departure.

In the UK, most procedural approaches have a beacon as the Initial Approach Fix (IAF). All you need to do is to enter in the IAF just prior to the destination airfield:

> SND
>
> EGMC

Now when you select the approach you still have a route from your present position to the IAF. Since you are likely to need to hold and/or enter the procedure at the IAF, including the IAF beacon in the route can be of immense benefit as the following pages will illustrate.

You can now proceed (subject to ATC) to the IAF to take up the hold or enter the procedure.

Another Look at the NDB

Let's assume you are entering the hold for an NDB DME approach to runway 06.

Go to the Active Route page and press **MENU** once to activate the context sensitive menu and choose **Select Approach**. This will display the available approaches for Southend. Select **NDB RW06**. If you have a GPSMAP 295 answer **No** to the question about Vectors.

As you approach the beacon you can prepare OBS mode to help you with the hold: With the SND NDB as the next active waypoint choose the HSI page, press **MENU** to activate the context sensitive menu and select **Set OBS and Hold**. The track/radial will appear on the HSI display. Use the rocker pad up/down to change the radial until it matches the inbound radial (053°) and press **ENTER**. You can now fly the hold using the ADF and timing as usual, however you now have the additional benefit that the inbound radial is shown on the GPS map and HSI pages.

When you are ready to commence the procedure you can use the context sensitive menu on the HSI page to **Release Hold** as you approach the beacon. Call "beacon outbound" and commence the procedure. You aren't following the GPS at this point, however the Active Route leg is now the leg from the FAF to the missed approach point so when you make the base turn you have the benefit of the inbound track on the map and the HSI.

Alert: You MUST use the proper aircraft DME for the procedure as the MAP or threshold which the GPS is measuring to is highly unlikely to be the same point as the DME in the procedure (and DME measures slant distance whilst the GPS measures horizontal distance).

Another Look at the ILS

Assume you are entering the hold for an ILS DME approach to runway 24.

Go to the Active Route page and press **MENU** once to activate the context sensitive menu and choose **Select Approach**. This will display the available approaches for Southend. Select **ILS RW24**. If you have a GPSMAP 295 answer **No** to the question about Vectors.

As you approach the beacon you can prepare OBS mode to help you with the hold: With the SND NDB as the next active waypoint choose the HSI page, press **MENU** to activate the context sensitive menu and select **Set OBS and Hold**. The track/radial will appear on the HSI display. Use the rocker pad up/down to change the radial until it matches the inbound radial (053°) and press **ENTER**. You can now fly the hold using the ADF and timing as usual, however you now have the additional benefit that the inbound track is shown on the map and the HSI.

When you are ready to commence the procedure you can use the context sensitive menu on the HSI page and select **Set OBS and Hold** again. Set the radial to match the outbound track of the procedure (043° in this case). This will give you situational awareness on the CDI as you travel away from SND.

As you reach the required distance outbound (6.5 DME) you can use the context sensitive menu on the HSI page to **Release Hold**. The route waypoints will start to sequence again and you should find that the last leg from the FAF to the runway threshold activates. At the same time you can commence the base turn. Of course in real life you will have identified the ILS and be using the localiser as your primary navigation, however you have the additional benefit that the inbound track is shown on the GPS Map and HSI pages.

If you have the GPSMAP 295 you can use the context sensitive menu on the HSI page and select Vectors to activate the final approach segment as an alternative to the last stage.

Summary

In this Chapter you've learned that there are quite a lot of complex operations needed in order to use the GPS for additional situational awareness during instrument approaches.

It is at your discretion how much of this you use. The Author prefers to fly instrument approaches using conventional radio navigation techniques, using the GPS map display for general situational awareness only.

5

Your GPS and Your PC

This section shows you how to update the Jeppesen database in your GPS and also how to upgrade the operating software.

Increasingly, pilots are able to do their flight planning using PC software and download waypoints and routes to and from their GPS receivers. The section covers three popular packages.

You will need to connect your GPS to your PC and this can make people nervous - as long as you are computer literate enough to plug your PC together and understand how to use files and folders in Microsoft Windows then it's really quite straightforward.

Alert: Garmin and the other vendors will no doubt continue to update and improve their software and web sites, and whilst this section is correct at the time of writing, it is possible that minor changes and discrepancies will creep in over time, however the basic principles should remain the same. For latest information see the web site that supports this book: **www.gpsbook.co.uk**

Introduction

This brief Chapter shows you how to connect your GPS to your PC. For further information, refer to your GPS manual and your PC manual if you have one.

Connecting

To connect your GPS you will need a PC connector cable. The GPS 92, GPS III Pilot and the GPSMAP 295 all use a similar cable with a four pin round plug. The GPSMAP 195 has a clip type connector. Only the GPSMAP 295 comes with the data cable as standard so if you have one of the other models you will have to acquire a suitable cable from one of the well-known aviation suppliers.

Tip: These cables used to be ridiculously expensive. They are now just expensive! It's well worth shopping around for the best price (and if you have a GPSMAP 195 make sure the supplier realises it uses a different cable type).

Before plugging or unplugging you should ideally turn off both your computer and the GPS.

Serial ports are also known as COM ports (COMmunication ports). Your PC may have more than one port and it is up to you which one to use. If you have only one port and already have a modem or other device connected to it you may need to disconnect it temporarily when you need to connect your GPS. These days most PC manufacturers annotate serial ports with 10101 and/or the COM port number.

One end of the data cable plugs into a 9 pin D connector serial port on your PC. If you have a 25 way D connector serial port you will also need a 9 pin to 25 pin converter (pictured in the middle above) which you can find at most PC and electrical stores.

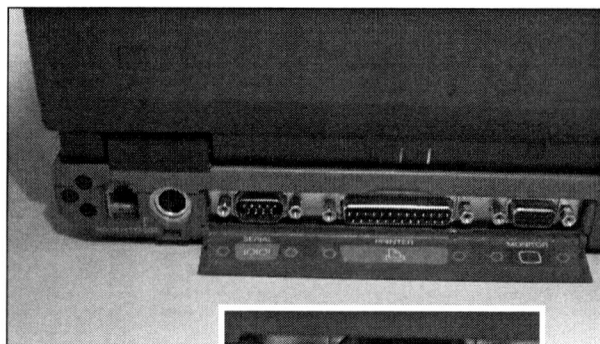

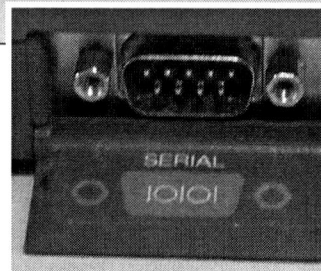

The other end of the cable plugs into your GPS. The GPS 92, GPS III Pilot and the GPSMAP 295 all use the round plug and you will find the socket under a rubber seal on the GPS. Ease the rubber seal out gently and plug in the cable, taking care not to bend the GPS pins. The GPSMAP 195 clip type plug clips around the body of the GPS and connects four pins to the connecter pad on the underside of the GPS.

Tip for GPSMAP 195 users: The Author has experienced difficulty getting a good connection due to the way the battery is moulded when using the GPSMAP 195 data cable. Using the Alkaline battery holder rather than NiCad improves the situation, and slipping a small piece of card between the battery and the GPS to displace it by about 1mm ensures a reliable data connection.

Introduction

Aeronautical information changes all the time so it is advisable to update the Jeppesen database periodically to make sure that it doesn't get too far out of date.

Fact: The database is published in 13 annual "cycles" each corresponding to about four weeks. The cycle is described by a four digit identifier where the first two digits are the year and the second two are the cycle number.

How often you update the database is partly a matter of choice and cost, however you should consider updating the database at least once a year. As this illustration of the Manchester low level corridor you transited in Chapter 15 shows, the effect of not updating the database for a couple of years can be quite significant!

Before   **After**

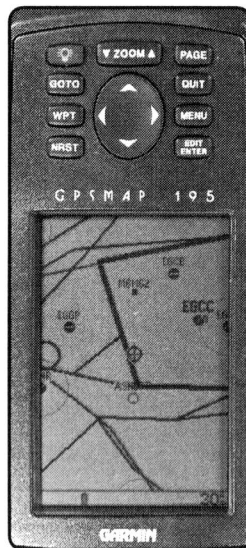   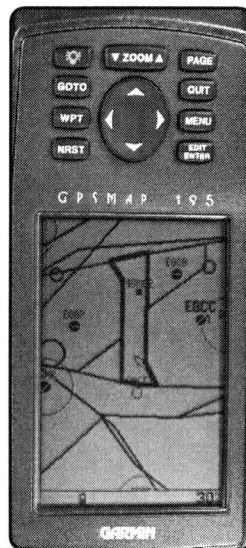

GPSMAP 195 and Manchester Corridor before and after a long overdue update.

Tip: The Author updates his database twice a year, once in spring around the time when the new aeronautical charts are published and again in autumn. He also updates it prior to a major overseas trip.

At the time of writing there are two ways of getting a database update:

- **Database update certificate.** You can buy the certificate from most of the popular aviation suppliers. Once you get the certificate you need to send it to Garmin Europe and they will send you back a couple of floppy disks containing the database and updater program.

- **The Garmin web site.** This method is rapidly superseding the certificate method. You can visit the Garmin web site at **www.garmin.com** select Aviation products and choose Aviation Databases from the sidebar. You can now choose the database update you require, download it, and pay online for an unlock key using a credit card. The download is keyed to your individual receiver. As well as getting individual updates you can subscribe for an entire year.

> **Tip:** At the time of writing getting a database update via certificate, mail order, can take around a fortnight and costs around £70 depending on supplier. Getting a database update online from the Garmin web site takes around 15 minutes and costs $35.

Before Updating your GPS

Connect the GPS to the PC as described at the start of this Section.

It is important that the Update process is not interrupted so you should make sure that there is reasonable battery charge in the GPS before you start.

The first part of the update erases the existing database so if the update process is interrupted you will find you have no database until you try again and successfully complete the procedure.

Using a Database Certificate

First acquire the certificate from one of the popular aviation suppliers in person or via mail order.

Fill out the certificate with the required details including your name and address, GPS model and the required database. Assuming you are flying in the UK and Europe, you will need the Atlantic International database.

Return the completed certificate to Garmin Europe (based in the UK) and they will send you the latest update within a few days. It will typically be on a couple of floppy disks including the database and the updater software.

Connect the GPS to the PC, switch them both on and restart the PC. When the PC has restarted insert disk 1 in the floppy drive and run the updater program. The program will first ask you to

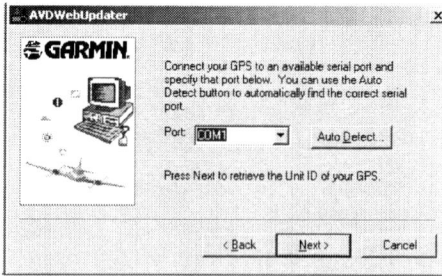

select the COM port the GPS is connected to. If you are unsure you should click the Auto Detect button and the program will scan the available ports. When the Port has been identified you can proceed. The program will verify the connection with the GPS and erase the existing database and then install the updated one. During the update process you will be prompted for any second disk if required.

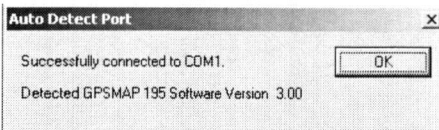

The actual update process typically takes around five to twelve minutes depending on the database size and the receiver model.

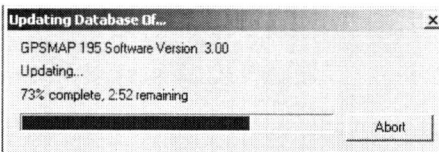

During the update it is possible that aeronautical waypoints have been deleted or moved. If you have any routes dependent on such waypoints you will receive a message that the Route Waypoint is "Locked". If this occurs it is recommended that you identify the route in question and re-edit it appropriately.

Updating Online

First go to the Garmin web site **www.garmin.com** and select the database update for your GPS model - at the time of writing, select Aviation Products and choose Aviation Databases from the sidebar. You will be shown an overview of the database options available and can continue on to see an overview of the steps involved.

Next you select your GPS model and then the database you need. Assuming you are flying in the UK and Europe, you will need the Atlantic International database. Next you need to agree the to the licence terms.

Fact: Amongst other things you will be agreeing to the terms that you will use the GPS only as a supplementary aid to VFR flights during the valid 28 day period of the database!

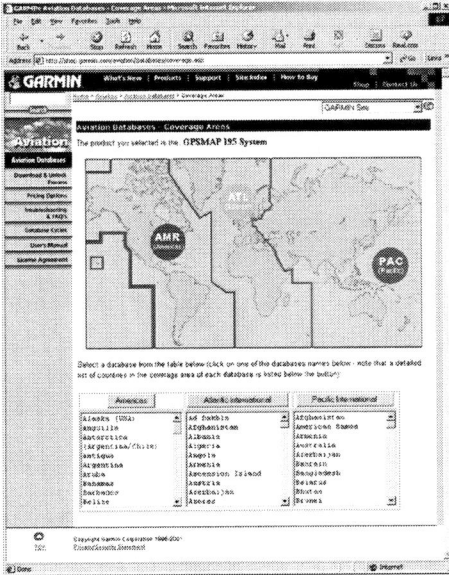

After doing this you can download the database updater program which takes less than 10 minutes on a typical modern modem connection. You should make sure you download it to a folder on your PC that you will be able to find easily when you need it again in a moment. The database file will include a code for the GPS model number, the database region (ATL in your case) and the four digit cycle number.

Next you need to make sure your GPS is connected to the PC as described earlier in this Section and switched on.

Alert: If you have an external modem on your only COM port (luckily quite rare these days), and this is how you are connected to the Internet, you will need to temporarily disconnect and juggle your connections between the GPS and the modem as required for the next steps.

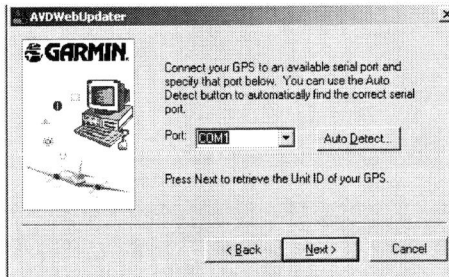

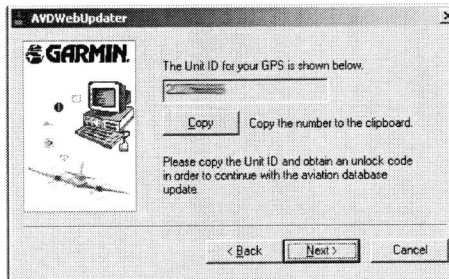

Run the program you downloaded. After the introductory screen, the program will ask you to select the COM port the GPS is connected to. If you are unsure you should click the Auto Detect button and the program will scan the available ports. When the port has been identified you can proceed. The Program will verify the connection with the GPS and display the receiver's "Unit ID". You will need to enter this ID into a form on the web site. You can copy it into the PC clipboard using the button provided and then paste it into the form (press CTRL-V).

You can now select the database option (whether you want a single update or annual service) and can enter in your credit card information. Provided this is satisfactory you will receive the unlock code and a password. You can use the password next time you return to the site to identify yourself. If you specified your email address earlier in the process this information will also be emailed to you.

Updating the GPS Database   **Updating Online**

Now you are ready to do the update itself:

Continue with the updater program (if you cancelled out of it after getting the Unit ID, just run it again and skip through the bits you have already done). You will be asked to enter the unlock key. When you have entered it satisfactorily the program will verify the connection with the GPS and erase the existing database and then install the updated one.

The actual update process typically takes around five to twelve minutes depending on the database size and the receiver model.

During the update it is possible that aeronautical waypoints have been deleted or moved. If you have any routes dependent on such waypoints you will receive a message that the Route Waypoint is "Locked". If this occurs it is recommended that you identify the route in question and re-edit it appropriately.

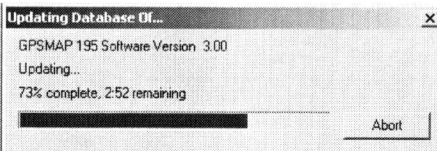

Trouble-Shooting

Updates will normally go very smoothly if you follow the instructions to the letter.

Here are some tips if you run into problems, and for further help you can consult the "readme" text files that come with the updater:

If the Updater program can't find the GPS:

Check the cable is plugged into the GPS properly (especially where the GPSMAP 195 is concerned)

- Check the cable is plugged into the COM port properly
- Check the correct COM port is selected (try auto-detecting if necessary)
- Check that the GPS data settings are correct (see next page)
- Does the COM port work with other devices?

If the Update fails:

- Check the unit has enough battery power to last for the length of the update.
- Try putting the unit into Simulator mode to prevent unwanted INIT messages.
- If it is a disk update and there is a disk fault you may need to get new disks from Garmin before retrying, though you can try copying to a hard disk or try the disks in another PC if you have access to one.
- If an online update fails, remember that each update is keyed to the specific receiver Unit ID - it won't work if you try to use the same update on a friend's GPS (not that you would do that of course!).

Checking the GPS Data Settings

By default, the standard GPS data (input/output) settings will work. While experimenting with your GPS you might have inadvertently changed them.

Checking the GPS Data Settings			
GPS 92	**GPS III Pilot**	**GPSMAP 195**	**GPSMAP 295**
From the Main Menu select **Setup Menu** and **Interface**. Check that the mode is set to **GRMN/GRMN**.	From the Main Menu select **Setup** and the **Interface tab**. Check that the format is set to **GARMIN** and the mode to **HOST**.	From the Main Menu select **Setup Menu** and **Input/Output**. Check that the Input/Output Format is set to **DATA TRANSFER** and the Transfer Mode to **Slave Mode**.	From the Main Menu select the **Interface** tab. Check that the Format is set to **GARMIN**.

Up-to-date Web Links

The web links described in this chapter may change as Garmin update their web site. If you have trouble, you can find up-to-date links to the correct place in the Garmin site at the web site that supports this book: **www.gpsbook.co.uk**

Introduction

Your Garmin GPS is actually a very powerful computing device. From time to time Garmin make improvements to its operating software. This is the software that makes your GPS work - it's a bit like the way Windows is the operating software that makes your PC work.

Garmin publish updates for download on their web site.

When you switch on your GPS it will show the software version number on the first screen it displays, except on the GPSMAP 295 that displays the software version on the System tab of the Main Menu.

Why Upgrade?

It's worth saying that it is not normally necessary to update the software unless you want to. However Garmin introduce new features and also fix bugs that may cause problems with the operation of your GPS.

Generally, if you have a GPS 92, you will have little cause to need to update your software. The GPSMAP 295 on the other hand has suffered from quite a few bugs including ones that made it corrupt routes and switch off spontaneously, and very early on it had a very cumbersome method of entering waypoints. The GPSMAP 295 also now supports WAAS and this was introduced with software version 2.20 (see Appendix A to find out more about WAAS). The GPS III Pilot and GPSMAP 195 may also need upgrading to work properly with some recent database updates.

The Updates & Download page for your GPS model will tell you the differences and benefits between each software version.

Before Upgrading the Software

Connect the GPS to the PC as described at the start of this Section.

It is absolutely critical that the Upgrade process is not interrupted, so you should make sure that there is good battery level in the GPS before you start. If you have a laptop PC you might wish to consider using it (connected to mains power) rather than a desktop PC, since if mains power is interrupted a laptop will continue operating on its battery.

> **Alert:** Only attempt this process if you are fully confident you can undertake it safely. If the update fails you may find that you are unable to operate your GPS and may need to return it to Garmin or an authorised Garmin reseller to have it reset.

Upgrading Online

First go to the Garmin web site **www.garmin.com** and find the update for your model of GPS - at the time of writing, select Aviation Products, Portable GPS and select your model. Next choose Updates & Downloads from the sidebar. You will be shown a list of the software available.

You should read the warning messages very carefully and then download the software. You should make sure you download it to a folder on your PC that you will be able to find easily when you need it again in a moment.

Next you need to make sure your GPS is connected to the PC and switched on.

The program you downloaded is a "self-extracting" compressed file that will expand itself to include an updater program, which will install the new operating software.

Once you have run the self-extracting file the updater program and the data file it uses will be in the folder.

The updater program varies between GPS versions. For example, the GPS 92 and GPSMAP 195 runs in a DOS window, whilst the GPS III Pilot and GPSMAP 295 uses Windows software. Make sure you read any readme text file or other information that comes with the update.

The actual update process typically takes around five to twelve minutes depending on the receiver model. When the process is complete, your GPS will restart automatically.

Trouble-Shooting

If you have problems you can use the same trouble shooting techniques described in the last Chapter.

If for any reason the upgrade fails and your GPS no longer functions, you will need to return it to Garmin Europe or an authorised reseller to be reset.

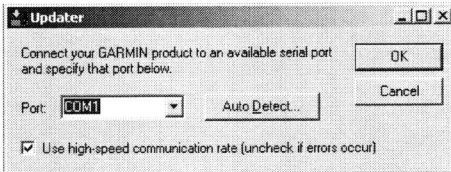

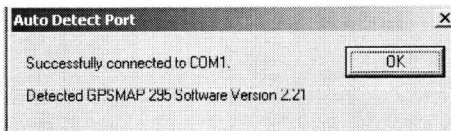

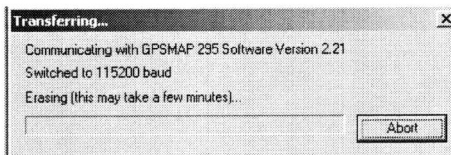

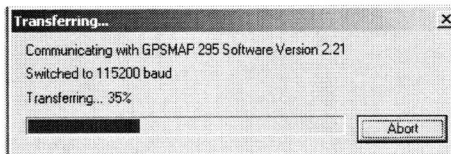

Introduction

Increasingly, pilots are able to do their flight planning using PC software and download waypoints and routes to and from their GPS receivers.

This Chapter looks at Garmin's own MapSource products, Jeppesen's Flitestar and NavBox's ProPlan.

This isn't intended to be a complete tutorial on the packages in question, rather an overview of the features. All the packages will recognise the Garmin Portable GPS receivers featured.

The Manufacturers of these products are constantly updating their software and exact specifications may vary over time. Its important to get the latest version as many of the better features are brand new at the time of writing.

Alert: Just like the GPS, these packages should be used as an aid to conventional navigation techniques. It's essential that if you use this kind of software for flight planning you still draw lines on a proper aeronautical chart and do a gross error check of the distances, angles and wind corrections against the flight logs generated.

Garmin MapSource

Garmin make available a number of products in the MapSource range. Not only do these products enable you to download detailed mapping into compatible GPS receivers such as the GPSMAP 295, but they also have route and waypoint management capabilities.

Having said this, the MapSource products are not really suitable for flight planning as they do not have a built in aviation database and they cannot extract the Jeppesen database from the GPS, only those points included in existing routes and user waypoints.

The mapping and additional information in the MapSource products can be very valuable to GPSMAP 295 owners when planning flying trips as they can give local road information and hotel data including phone numbers.

Unfortunately MapSource is let down by the fact that the CDs are quite expensive and the data cartridges equally so. They are also limited by how much map data you can fit in them and this can defeat the benefit for a trip covering a lot of ground. The GPS receivers also use serial data transfers (rather than more modern links like USB or "Firewire") and so downloading the maps to the GPS can take quite a long time.

A Garmin MapSource Cartridge

Jeppesen FliteStar

Jeppesen make a range of very professional products under the FliteStar banner including VFR, IFR and commercial versions. This book focuses on the VFR package, as it is the more affordable version (The IFR version costs around three times as much).

This software is very comprehensive and easy to use and is customisable in many ways, however like many Jeppesen products it omits ATZs, though MATZ information has recently been introduced in version 8.4 (the latest version at the time of writing).

Older versions of the software had the disadvantage that when you uploaded a route to the GPS it added all the

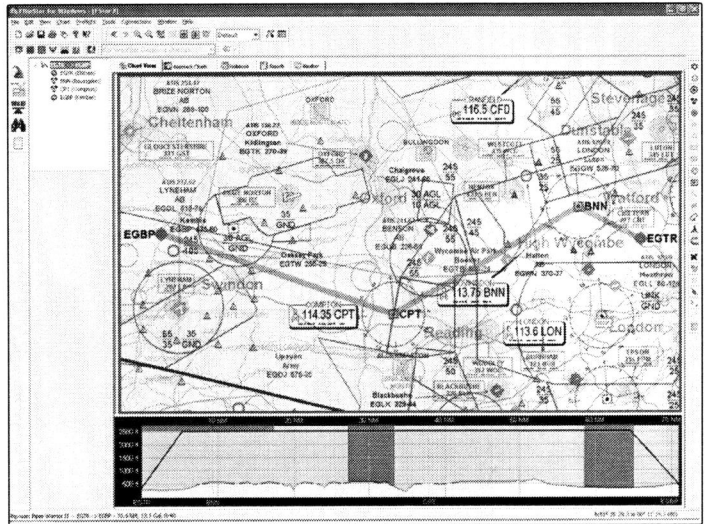

The Elstree to Kemble route from Chapter 11 planned using Jeppesen FliteStar

points as user defined waypoints even if they were aviation waypoints already in the GPS (Jeppesen marketed this as a feature claiming that it meant you could use non-aviation GPS receivers for aviation work).

The latest version of FliteStar has an option to allow you to use the aviation waypoints in an aviation GPS if they are present, however this only seems to work properly on the GPS III Pilot and GPSMAP 295. The GPS 92 still creates unnecessary user waypoints and the GPSMAP 195 is OK unless the route has user waypoints in it (in which case they are transferred to the GPS but deleted from the route!).

FliteStar can't be used as a moving map though a derivative called FliteMap can, though at some considerable additional expense.

NavBox ProPlan

This is the cheapest of the three packages, and arguably the best.

When uploading routes to the GPS the program will use aviation waypoints present in the GPS and will automatically upload any waypoints that don't exist in the GPS database as user waypoints.

A big advantage is that it includes waypoints of use to VFR pilots such as non-ICAO airfields, such as Ashcroft used as an example earlier in this book.

Version 4 of the product, which has just been introduced at the time of writing, can be used as a moving map (if you can fit a laptop in the cockpit with you!).

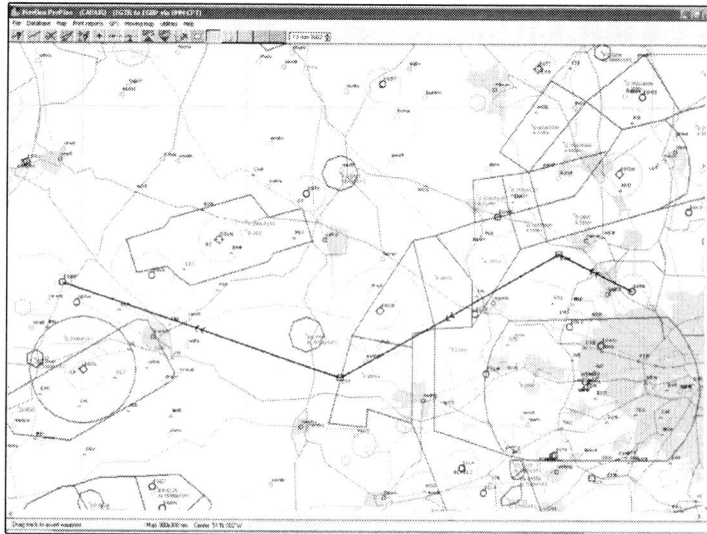

The Elstree to Kemble route from Chapter 11 planned using NavBox ProPlan

The only slight problem with NavBox is that zooming and scrolling the map is slightly cumbersome.

Database Updates are available to subscribers at **ww.navbox.nl** and **www.avnet.co.uk/navbox**

Up-to-date Web Links

The web links described in this chapter may change as companies update their web sites. If you have trouble, you can find up-to-date links at the web site that supports this book: **www.gpsbook.co.uk**

Summary of Features
General Functions

Feature	MapSource	Flitestar	NavBox
Visual planning on map	✓	✓	✓
Planning as text	✗	✓	✗
Weather planning	✗	✓	✓
Live Weather by modem/Internet	✗	✓	✗
Flight Logs	✗	✓	✓
ICAO flight plan	✗	✓	✓
Aircraft models	✗	✓	✓
Weight and Balance	✗	✓	✓
Automatic Frequency planning	✗	✗	✓
Chart printing	✗	✓	✓
Moving Map	✗	✗ *	✓

*Jeppesen do a product called FliteMap that is significatly more expensive than Flitestar and provides moving map and in-flight facilities.

Built-in Waypoints

Feature	Garmin GPSMAP	MapSource	Flitestar	NavBox
Airfields (ICAO)	✓	✗	✓	✓
Airfields (non ICAO)	✗	✗	✗	✓
Heliports	✗	✗	✗	✓
VOR	✓	✗	✓	✓
NDB	✓	✗	✓	✓
DME	✗	✗	✗	✓
TACAN	✓	✗	✓	✓
Intersections	✓	✗	✗*	✓
Approaches	✓	✗	✗*	✗
Departures	✗	✗	✗*	✗
VRP	✗	✗	✓	✓
Towns (visible on map)	✓	✓	✓	✓
Towns (usable as waypoint)	✓	✓	✗	✓

*Available on the IFR version of Jeppesen Flitestar

Airspace

Feature	Garmin GPSMAP	MapSource	FliteStar	NavBox
ATZ	✗	✗	✗	✓
CTR	✓	✗	✓	✓
CTA	✓	✗	✓	✓
Airways	✗	✗	✗*	✓
MATZ	✓	✗	✓	✓
AAIS	✓	✗	✓	✗
Restricted	✓	✗	✓	✓
Danger	✓	✗	✓	✓
HIRTA	✗	✗	✗	✗

*Available on the IFR version of the Jeppesen FliteStar.

GPS Functions

Feature	MapSource	FliteStar	NavBox
Download Routes from GPS	✓	✓	✗
Upload Routes to GPS	✓	✓*	✓
Download User Waypoints from GPS	✓	✓	✗
Upload User Waypoints to GPS	✓	✗*	✗*
Download Track Log from GPS	✓	✓	✓
Download Maps from GPS	✓	✗	✗
Upload Maps to GPS	✓	✗	✗

*User defined waypoints required as part of a route are uploaded.

Other Software

There is also quite a lot of older software available as well as shareware. This software often isn't very good and sometimes suffers from the same problem that the FliteStar software has had, which is to create new waypoints oblivious of the inbuilt aviation database in the GPS.

6

Introducing the GPSMAP 196

Introduction

The GPSMAP 196 was introduced in August 2002.

This Section introduces this new GPS and highlights the key differences with the other units.

The GPSMAP 196 interface shares a lot of features in common with the GPSMAP 295 (though, because it is somewhat older, less so the GPSMAP 195).

Although the GPS has been designed for aviation, land and marine use, this Section focuses on aviation use.

Powering the GPS

The GPS can be powered by four AA batteries or supplied by external power. Batteries will power the unit for around 16 hours, or 4 hours if backlighting is used. Power is discussed in more detail in Chapter 3.

Switching on the GPS

Switch on the GPS by pressing and holding the red power button briefly. The unit will go through a similar power-up sequence to the one described in Chapter 4.

The GPS will try and auto-locate itself, however if it has problems (for example if you are trying it out indoors or without an antenna attached) it will alert you with a warning menu. Garmin have made this more user-friendly than on earlier models and amongst the options ask if you are at a new location or if the unit has been stored without batteries. This is synonymous with initialising your position or auto-locating on earlier models.

Power button

Familiarising yourself with your GPS

Like the other GPS units featured in this book, the GPSMAP 196 has a "main page loop", which are the screens used most often in the air, functions accessed directly from buttons and a Main Menu. These are similar to the features described in Chapter 5.

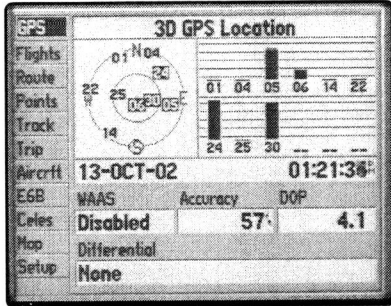

The Satellite Status Page is displayed alongside the Main Menu.

The Main Page Loop on the GPSMAP 196 is very similar to other Garmin GPS portables and consists of the following pages:

- Moving Map Page
- HSI Page
- Active Route Page
- Position Page

To move forward between these pages, press the PAGE button and to move backwards, press the QUIT button.

Like the GPSMAP 295 screens these can be configured to change the number and size of displayed fields and also the layout. The HSI screen can be configured to act as the simulated aircraft panel featured in many adverts for the GPSMAP 196.

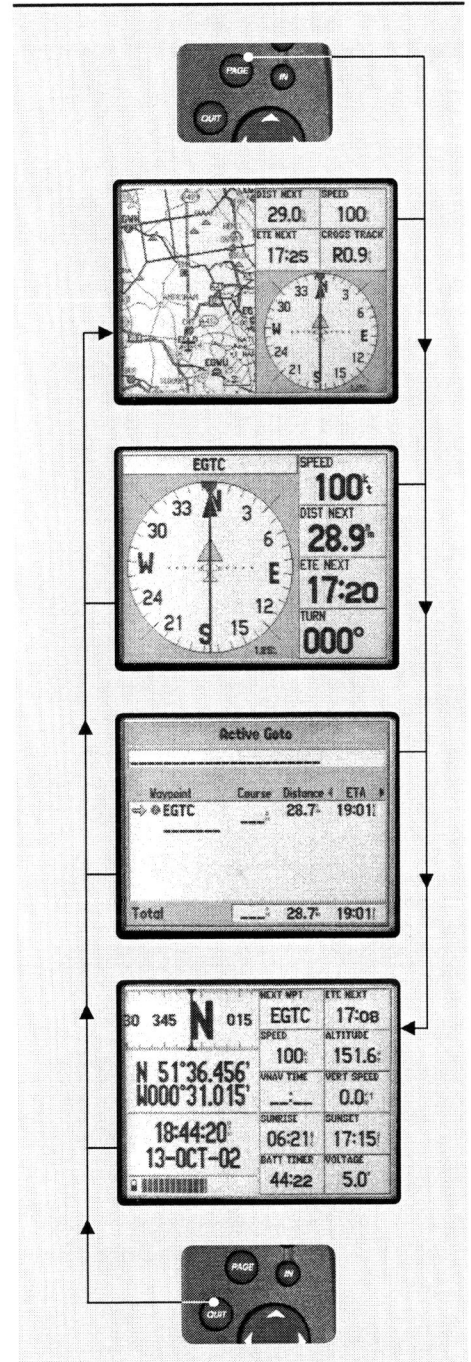

The GPSMAP 196 Moving Map Page

Unlike earlier receivers, the map is considerably more readable at its default settings at levels of zoom likely to be used in the UK and Europe, and a Declutter menu option on the Map screen (or pressing the ENTER MARK button quickly) let's you declutter the map in stages.

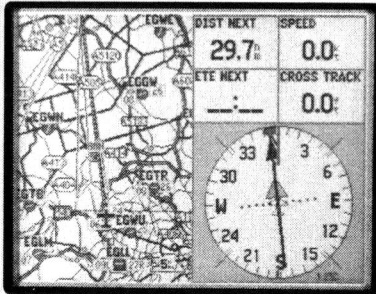

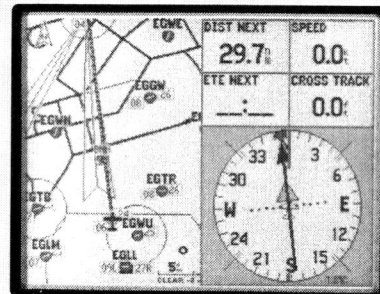

| Normal | Partly Decluttered |

There are couple of additional options in aviation mode:

- Heading Line extrapolates the aircraft's present ground track and shows a line ahead of the aircraft symbol indicating its path (similar to the feature on Skymap GPS systems). By default this is not enabled.

- Rnwy Extensions shows "arrow fleche" annotations extending out from the destination airport's runways.

- Bearing Line is a new feature introduced in a recent system update (so not available on early models), which extrapolates the bearing to the next waypoint.

The Map setup can be changed using the Map tab on the Main menu screen or by pressing MENU to activate the context sensitive menu and selecting Setup Map. Most of the settings are similar to those on other Garmin receivers (see Chapters 9,13 and Appendix B).

One criticism of the otherwise excellent map display is that the lines used to depict control zones are so thin as to be very difficult to see during flight. Other controlled airspace is easily visible (though of course, like other Jeppesen VFR products, the airspace included does not include airways).

The GPSMAP 196 Simulated Aircraft Panel Page

To activate this screen, go to the HSI page, press ⊙ to bring up the context sensitive menu, select Setup Page Layout and select Panel.

The first time you go to the panel screen after activating it will display a warning that the Panel is for VFR use and the display may differ from the aircraft instruments.

It is reasonable to argue that this screen is a gimmick. The displays only approximate to the aircraft instruments:

- The ASI equivalent measures groundspeed not airspeed

- The Altimeter is inaccurate (see Chapter 2 for a discussion of GPS accuracy)

- The HSI measures ground track, not magnetic heading (see Chapter 13)

The turn coordinator is the most unreliable. It computes its display from perceived turns. In practice, even in light turbulence this is as good as useless and it has frequently been seen to even show turns opposite to the real turn coordinator.

GPS Buttons and the Main Menu

Broadly Speaking the GPSMAP 196 buttons operate in a similar way to the GPSMAP 295 described in Chapter 6, though unlike the 295, the 196 doesn't have a Route button - Instead the routes are accessed from the Main Menu.

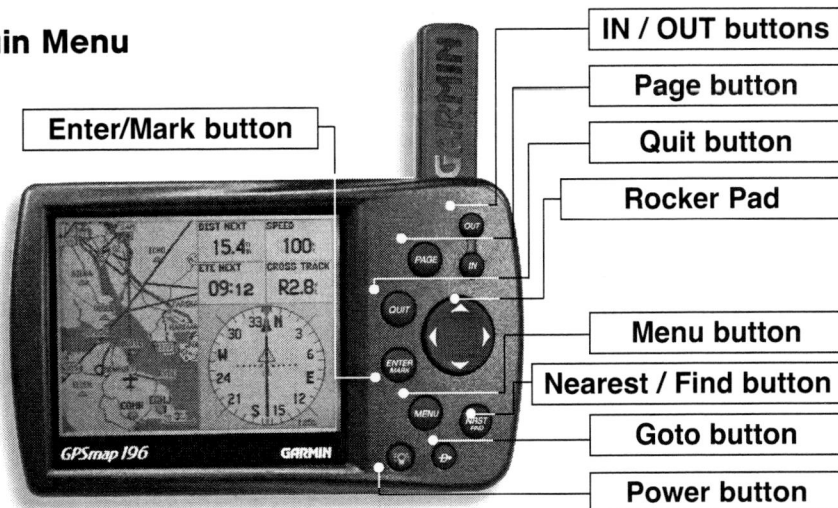

Enter/Mark button

IN / OUT buttons

Page button

Quit button

Rocker Pad

Menu button

Nearest / Find button

Goto button

Power button

To access the Main Menu, press the ● button twice. The Main menu is displayed as a set of vertical tabs along the left hand side of the screen. To select different sub-menus use up/down on the rocker pad.

The GPS tab shows the Satellite Status Screen and also enables WAAS and Differential GPS to be enabled. Pressing ● to display the context sensitive menu lets you start and configure the Simulator.

The Flights tab shows a logbook style presentation. Recording of flights starts when you climb 500 feet and exceed 30 knots. The log does not record takeoff and landing time but only flight duration and is easily deleted so it is not really a substitute for proper logging of flights though is useful if you forget to log your takeoff time.

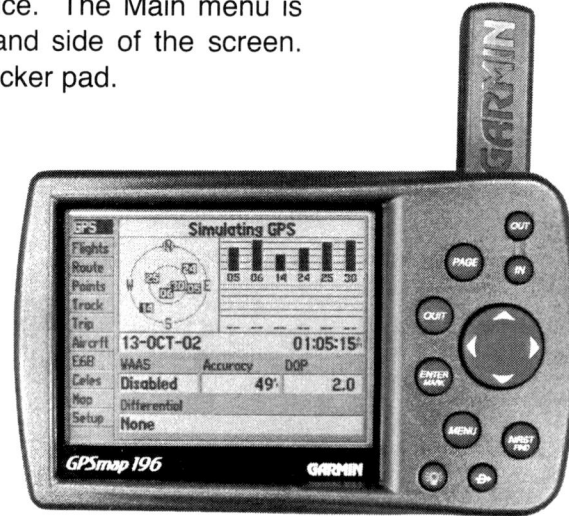

The Route tab enables you to enter routes (See Chapter 13). You can enter a new route by pressing ● to display the context sensitive submenu and selecting New Route or simply selecting the blank route and pressing the ● button. Garmin have increased the capacity of the GPSMAP 196 and it can store 50 reversible routes of up to 50 points each

The Points tab displays user waypoints and pressing ● to display the context sensitive submenu provides options for creating and managing them. You can enter up to 1000 user waypoints. The Proximity sub-tab is a new feature which enables you to set proximity alarms around waypoints.

The Track tab enables you to manage and store up to 15 track logs, enabling you save previous tracks. This feature is perhaps more useful in marine mode however can be useful for easily representing "scenic routes". You can also use these track logs in connection with the TracBack feature (see Chapter 17).

The Trip tab displays odometer functions including totals, averages and trip information for distances, times and speeds.

The Aircrft tab lets you input aircraft information including call signs, performance and fuel flow. The Weight & Balance sub-tab enables you to enter a weight and balance schedule for the chosen aircraft. You can only enter one aircraft at a time.

The E6B tab contains an electronic version of features that can be calculated using an E6B circular slide rule including True Airspeed and density altitude and winds aloft.

The Celes tab shows sunrise and sunset times, though perhaps is otherwise most useful for marine users.

The Map tab enables you to set up the map to meet your preferences. As mentioned above, most of the settings are similar to those on other receivers (see Chapters 9,13 and Appendix B).

The Setup tab enables you to access a series of sub-tabs to configure the GPS. This includes the following features:

- System, including Simulator and usage mode, backlight and beeper settings and language
- VNAV Profile enables you to pre-program descent warnings (see Chapter 13)
- Airspace alarm setup, enabling you to configure airspace warnings (see Chapter 9)
- General Alarms, enabling you to setup various warnings and time based alarms.
- Road Routing, enabling you to set up routing options for land use.
- Timers, which as well as timers found on previous models now includes the ability to set up fuel tank change times.
- Time, including time zone and format configuration options (see Chapter 8)
- Units, including pressure, temperature, speed and altitude configuration (see Chapter 8)
- Location, including format map datum and variation set up (see Chapter 8)
- Interface, for configuring the GPS for PC connection.

Installing the GPS in the aircraft

The GPSMAP 196 comes with yoke mount and glare shield mount options. The glare shield mount can also be used in a car or boat.

It is best to yoke mount the receiver if possible as its larger screen seems to make it more susceptible to glare than other units. Unfortunately, like the GPSMAP 295, it is just slightly too wide to fit between the horns of newer Piper yokes and no doubt other aircraft.

The standard antenna may be adequate in many circumstances, but for best results use the included external antenna.

Install the GPS using the guidelines given in Chapter 10. The GPSMAP 196 used in the writing of this book had a rather dark screen, and the brightness had to be increased for convenient operation.

Flying with the GPSMAP 196

The GPSMAP 196, like all the Garmin receivers, is a tool for aiding navigation and situational awareness and should always be used in conjunction with proper flight planning.

Sections 3 and 4 of this book describe these techniques in detail. The GPSMAP 196 has all the features of the GPSMAP 295 and more (except the colour screen) and generally the functionality is similar.

Upgrading the Software and Database on the GPSMAP 196

The GPSMAP 196 comes complete with a PC cable and can easily be connected as described in Chapter 21.

Like the earlier models of GPS you can upgrade the Jeppesen database and the operating software. The GPS also comes with a voucher with a code for a free Jeppesen update via the web.

At the time of writing there are already several operating system upgrades available fixing quite a few bugs and improving a number of features, so it is probably worth considering an update, especially if you own an early unit.

The GPS can be updated using the techniques described in Chapter 22.

The GPSMAP 196 and Flight Navigation Software

Like the earlier Garmin units, the GPSMAP 196 can be used with flight planning software such as Jeppesen FliteStar and NavBox ProPlan. It can also be used in conjunction with a MapSource Cartridge and CD-ROM.

At the time of writing there appear to be a couple of issues concerned with correct operation of the GPS with this software, even though the software works satisfactorily with both the GPS Pilot III and the GPSMAP 295.

Tests have shown that both FliteStar and NavBox packages will upload all waypoints as user defined waypoints into the GPS (in the former's case, even with the "use GPS waypoints" option checked). This means that you get square blobs all over the proper aviation symbols on the map and rapidly use up your user defined waypoint list.

FliteStar also seemed to be unable to download routes from the GPS without corrupting them.

NavBox may need to be upgraded before it will recognise the GPSMAP 196, however for web users this upgrade is a simple, quick, free download from www.navbox.nl.

It is quite probable that these problems will be resolved with future upgrades to the GPSMAP 196 operating system and/or the flight planning software.

7

Garmin GNS 430 Supplement

The Garmin GNS 430 is becoming increasingly popular. As well as providing a moving map GPS it includes COM and NAV radio functionality, and many flying organisations are using it as a replacement for older non-FM immune equipment.

This supplemental section is aimed at readers who rent or hire aircraft with a single or twin GNS 430 installation on board and is designed to build on the knowledge contained earlier in this book to help with the basic operation of these units.

The manuals that come with the Garmin panel-mount receivers are very comprehensive - more so than the portable receivers. This section isn't intended to replace those manuals - Chapter 26 shows you how to get hold of your own copy of the manual from the Internet.

Many of the principles in this section can also be applied to the larger GNS 530 and Garmin's older or less sophisticated panel mount GPS systems.

Introduction

The Garmin GNS 430 is a very popular panel mount unit, which includes the functionality of a COM radio, NAV radio and moving map GPS.

The GNS 430 is TSO-C129 compliant, which means that in some territories it will be possible to use it as the primary flight navigation system (even including approaches). At the time of writing, no GPS systems may be used for primary flight navigation in the UK.

The GNS 430 has considerably more functionality than any of the portable GPS receivers discussed in this book, though the map features are arguably similar to the GPSMAP 295.

Manuals and Reference Guides

The GNS 430 comes with a very comprehensive manual and a quick reference guide. Both are quite substantial and even the quick reference is not ideally suited to in-flight use, and so a thorough understanding of the basic functionality is essential.

If you hire the aircraft you fly then you may not be able to easily borrow these documents. If this is the case then you can download the main manual from the Garmin web site:

At the time of writing, select Aviation Products, Panel-Mount GPS and select GNS 430. Next choose User's Manual from the sidebar. The download is fairly substantial and will take around 20 minutes using a modern modem. The manual comes as a "PDF" file and so you will need Adobe Acrobat if you haven't already got it - all this is made clear on the web site.

The GNS 430 Simulator

Learning how to use the GNS 430 at home would be almost impossible if it wasn't for an excellent simulator available from Garmin. Unlike the in-built Simulators on the portable receivers, the GNS 430 Simulator runs on a PC. Garmin calls the Simulator the "Series Trainer" and it also covers the other models in the 400 series range.

The Simulator can be downloaded from the Garmin web site. At the time of writing, select Aviation Products, Panel-Mount GPS, select GNS 430 and look for the Simulator download. The Simulator comes as a "Zip" file and after you download it you will have to first "unzip" it using a utility that recognises the zip format such as PKZip for Windows. This extracts all the installation files. Next install the program by using the Setup file you have just extracted and following the instructions.

Tip: Zip files were originally invented by a company called PKWare and are used very commonly to transfer data from the Internet. Using Zip files is beyond the scope of this book. For more information and to download a free evaluation copy of PKWare for Windows see **www.pkware.com**

The Simulator also features a cockpit HSI and has autopilot style controls.

When you first install the Simulator it defaults to the Americas database. You can select the Atlantic International Database (which includes the UK and Europe) by going to the Options Menu and selecting System Setup and changing the Jeppesen Database to "intrnatl.img" and clicking OK. Once you've done this you will need to turn the GPS back on by clicking on the simulated COM/Power/Volume knob (.C).

You can also initialise the default position of the Simulator to your preferred airfield using the Initialize **Position** option from the **Options Menu** and using the right hand outer and inner knobs to enter the appropriate ICAO code.

To get instant help about the simulator press F1.

Introduction

This Chapter tells you everything from turning the GPS on to the buttons, knobs and main pages of the GNS 430 and how to find your way around them. This is just an overview. You may wish to refer to the GNS 430 manual and the earlier Chapters in this book for more information.

Switching the GNS 430 On

To switch the Unit on, turn the [.C] knob clockwise. The GNS 430 will go through its self-test sequence and display the valid dates of the Jeppesen Database. To acknowledge the database date information, press the [ENT] button.

Alert: It is very common to find rental organisations have not updated the database since the GPS was originally installed! The problems of flying with an out of date database have been highlighted throughout this book. The GNS 430 can only be used as a supplemental navigation aid in the UK, and in those territories where it is certified for use as a primary navigation aid the certification is normally invalidated if the database is not current.

Checklist: You should switch on the GPS at the same point that you switch on the other avionics and ideally check the status of external navigation gauges such as the CDI (or HSI) during the Instrument Panel Self-Test.

Next, depending on the installation, the Instrument Panel Self-Test page may be displayed. The GNS 430 can be connected to external instruments including CDI (or HSI), altimeters, weather and traffic detection systems, fuel sensors and so on.

This page can also be configured to help manage fuel and display certain checklist items. The detail of this is beyond the scope of this book.

Next the GPS will start acquiring satellites and display the Satellite Status page. This is normally very rapid, though if the GPS has not been used for a few months or the aircraft has been relocated while the GPS was switched off, it may display "Searching the Sky". In this case satellite acquisition may take five to ten minutes.

If the GPS cannot receive a good signal, the word INTEG (for Signal Integrity) will show yellow towards the bottom left of the screen.

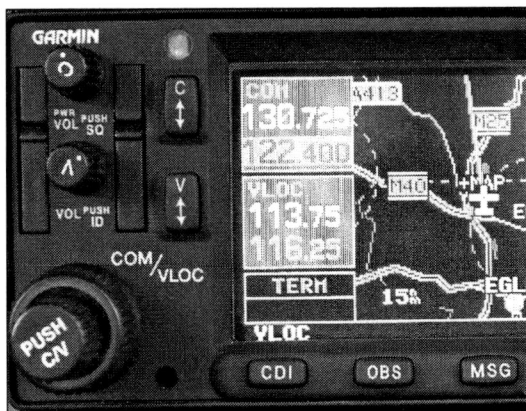

COM and NAV Buttons

The buttons and knobs at the left hand side of the GPS are similar to the equivalent buttons found on many conventional COM and NAV radios.

The [.C] knob is the COM volume and squelch knob. It also is the main power knob for the whole unit. Turn the knob to adjust the COM volume and press it briefly to toggle the automatic squelch on or off.

The [C flip flop] button switches the standby and active COM frequencies displayed in the "fields" adjacent to the button.

Tip: Press and hold the [C flip flop] button to tune immediately to the emergency frequency (121.5).

The [.V] knob is the volume knob for the selected NAV frequency. Garmin call this the VLOC volume control, standing for VOR/Localiser. Press it briefly to toggle the ident tone on or off.

The [V flip flop] button switches the standby and active VLOC frequencies displayed in the fields adjacent to the button.

The inner [PUSH C/V] knob tunes the KHz value of the standby COM or VLOC frequency. By default the COM frequency is tuned. Press the knob briefly to toggle the tuning cursor between the COM and VLOC fields. The cursor automatically returns to the COM field after 30 seconds.

The outer [PUSH C/V] knob tunes the MHz value of the standby COM or VLOC frequency.

Main GPS Buttons

The buttons and knobs at the right hand side of the GPS control the main GPS functions and are similar in many ways to the buttons found on the portable GPS receivers.

The [RNG] button enables you to zoom in and out when a map page is displayed. Use the down arrow side to zoom in and the up arrow side to zoom out.

The [Goto] button gives you instant access to the Goto function, which is one of the easiest to use and most powerful navigation features of the GPS. On the panel

mount GPS receivers, Garmin call this the "Direct-to" function.

The [Menu] button selects a context sensitive menu of commands relevant to the page you are on. Unlike the portable receivers, the GNS 430 doesn't have a "Main Menu" so pressing the button a second time simply cancels the menu displayed.

The [CLR] button is used to cancel an entry or delete information.

> **Tip:** Press and hold the [CLR] button to get back to the Default Navigation Page from whichever page you are on.

The [ENT] button is used to confirm a selection or entry.

The inner [PUSH CRSR] knob works differently depending on whether or not there is a cursor on the screen. Press the knob briefly to toggle the cursor on or off. When there is no cursor, turning the knob selects specific pages (or "screens") of information from within a main page group. When the cursor is enabled it makes selections for example to change and select letters to specify ICAO identifiers etc.

The outer [PUSH CRSR] knob works differently depending on whether or not there is a cursor on the screen. When there is no cursor, turning the knob selects which of the four main page groups is displayed. When there is a cursor the knob enables you to move the cursor between fields.

> **Tip:** In many ways the right hand knobs perform similar functions to the rocker pad found on Garmin's portable GPS receivers. The inner knob is similar to the rocker pad up/down and the outer knob to the rocker pad left/right

Bottom Row Buttons

Beneath the display are a number of additional buttons that give direct access to key GPS features:

The [CDI] Button toggles whether GPS or VLOC information is relayed to the cockpit CDI or HSI. The mode is displayed on the screen just above the key. In GPS mode, the external display will use the GPS information and in VLOC mode the external display will use information relative to the tuned VLOC frequency.

VLOC mode: To use the GNS 430 as a conventional NAV radio with an external CDI or HSI, use the [CDI] button to make sure it is in VLOC mode.

The [OBS] Button toggles OBS mode on and off. When you select OBS mode, the GNS 430 will normally take the OBS setting from the setting of an external CDI (or HSI). In some installations you may be prompted to set the OBS radial manually on the GPS instead (use the right hand knobs). Some uses of OBS mode were discussed earlier in Chapter 19. OBS mode is also used on the GNS 430 in connection with sequencing approaches. When OBS mode is active, this is indicated on the screen just above the button.

The [MSG] Button displays any active messages. A MSG alert on the screen just above the button flashes yellow when a new message occurs.

The [FPL] Button takes you directly to the Flight Plan pages. Flight Plan pages are similar to the Route pages on the portable GPS receivers and enable you to create and manipulate routes.

The [PROC] Button enables you to directly select, activate or remove approaches, arrivals and departures for your destination.

Main GPS Pages

The GNS 430 has four groups of main pages - NAV, WPT, AUX and NRST. The page group and the page selected is indicated at the bottom right of the screen. To select a page group use the outer [PUSH CRSR] knob, and to select the exact page within the page group use the inner [PUSH CRSR] knob.

NAV Pages

The NAV page set normally consists of six navigation pages:

- **Default Navigation Page.** This contains a CDI style display and key information about the current leg of your route.

- **Moving Map Page.** This displays the GPS Map screen. You can press [Menu] and use the context sensitive menu to change the map configuration. The points discussed in Chapter 9 should be considered in this context too.

- **NAVCOM Page.** This displays information about the frequencies of airfields on your route. You can change the airport displayed by activating the cursor and editing it. If you activate the cursor and highlight any frequency displayed and press [ENT] it will automatically be stored in the standby COM frequency.

- **Position Page.** This displays the current position (latitude and longitude), a compass heading and other useful information including the time.

- **Satellite Status Page.** This displays the satellite status including which satellites are being received and estimated position accuracy. You should check this page when you first turn the GPS on and during your cruise checks.

- **Vertical Navigation.** This page enables you to configure vertical navigation parameters. An example of using Vertical Navigation appears in Chapter 13.

On the majority of these pages you can press [Menu] and use the context sensitive menu to edit the fields. The default fields seem to be more useful than on some of the portable receivers, though you may wish to consider changing ETE to ETA on the Default Navigation Page.

The Garmin GNS 430 can have other devices connected to it including Weather and Traffic detection systems. The data from these devices will normally be displayed on a seventh NAV page.

WPT Pages

The WPT group consists of ten pages for waypoint information.

The first few pages are in common with the portable receivers: there are the three main airport information pages including location, runway and frequency information, plus pages for intersections, NDBs, VORs and user waypoints.

In addition there are three additional airport pages that show Approaches, Arrivals and Departures.

It's well worth experimenting with these pages using the Simulator to become conversant with all the information contained in them.

Enabling the cursor on any waypoint page, highlighting a frequency and pressing [ENT] will load that frequency into the standby COM or NAV field as applicable.

User waypoints can be defined in terms of latitude and longitude or as a relative bearing and distance from a pre-existing waypoint. Press [Menu] on the user waypoint page in order to list, edit and delete user waypoints.

AUX Pages

The AUX group consists of four pages, which are essentially menus of auxiliary functions and configuration options. To select an item from an AUX page enable the cursor by pressing the inner [PUSH CRSR] knob and use the outer [PUSH CRSR] knob to select the desired menu option and press [ENT].

Flight Planning Page

Fuel Planning. This enables you to compute fuel required between two waypoints or for a route (flight plan) based on manually entered fuel flow information. Fuel information will be entered automatically if appropriate sensors are fitted to the aircraft.

Trip Planning. This enables you to look at distance and time estimate information between two waypoints or for a route (flight plan), based on planned speed.

Density Alt / TAS / Winds. This enables you to use the "E6B" functions of the GPS. This is similar to the functions described in Chapter 17.

Crossfill. This feature is only enabled in a dual installation. The settings determine the inter-relationship between the units - this is described in detail in the next Chapter.

Scheduler. This option is of most use for aircraft owners and enables them to set up reminders for scheduled servicing of equipment etc. It can also be used to set up in-flight messages (such as prompts to switch fuel tanks etc).

Utility Page

Checklists. This option is of most use for aircraft owners and operators, and enables them to set up and use checklists for various stages of the aircraft operation. Use of these checklists is deliberately beyond the scope of this book, as there is no guarantee they have been set up and maintained correctly!

Flight Timers. This provides flight timers similar to those available on the portable GPS receivers.

Trip Statistics. This provides odometer and other trip information similar to those available on the portable GPS receivers.

RAIM Prediction. This feature predicts if GPS satellite reception will be sufficient for a planned flight to ensure approach quality coverage. Since GPS systems are unapproved for approaches in the UK this feature is presently little used.

Sunrise/Sunset. Predicts the sunrise and sunset at a given waypoint or position.

Setup Pages

There are two setup pages which provide facilities similar to the setup menu on the portable GPS receivers including airspace alarms, units, position datum, date/time, nearest airport criteria, display illumination, plus COM settings.

NRST Pages

The NRST group consists of eight pages, each displaying the nearest object in a different information category.

You can display the nearest Airports, Intersections, NDBs, VORs and User Waypoints as well as the Nearest Airspace.

You can also display the nearest Center and Flight Service information. In the UK these approximate to nearby Flight Information Region and Air Traffic Service frequencies respectively, though seem to give quite unexpected results.

Enabling the cursor on any NRST page, highlighting a frequency and pressing [ENT] will load that frequency into the standby COM or NAV field as applicable.

Introduction

This Chapter is designed to give quick advice if you are hiring a plane with a GNS 430 (or pair of GNS 430s). It builds on the background information from the past couple of Chapters and the rest of the book to highlight the safe usage of key features.

> **Airmanship:** Check out the GPS on the ground before flight.

Switching On

The GNS 430 is switched on by rotating the [.C] on/off/volume knob clockwise.

The GPS will sequence through similar pages to the portable receivers, displaying a "title" screen during which it does a self-test and then displaying the Aviation database details. Sadly, in hire aircraft it is common for the database to be significantly out of date. Although the database is likely to be called an IFR database, in the UK, the GPS is only valid as a supplementary navigation aid. Even in the United States, the GPS is not valid for IFR use if the database is out of date.

Next the Instrument Panel Self-Test will appear. The GNS 430 will drive any connected CDI/HSI and RMI to the settings indicated on the screen - check the external instruments against the settings shown.

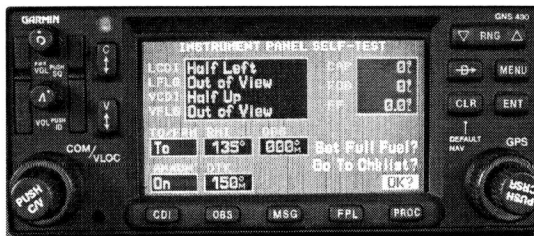

GPS Display with connected CDI showing localiser and glide slope needles driven left and up and the TO flag enabled.

> **Checklist:** Make a point of checking the database and Instrument self-test at an appropriate point in your checklist - for example when switching the avionics on.

The self-test page also may proffer fuel and checklist options. Unless you are very familiar with the particular aircraft and know the configuration of these options to be correct it is suggested you use conventional fuel management and checklists.

Subsequently the satellite status page will be displayed. A yellow INTEG field will be illuminated towards the bottom left of the screen when satellite reception is poor

Getting to a Common Starting Point

If you get lost within the complexities of menus and options of the GPS, pressing and holding the [CLR] key will return you to the main default NAV page from wherever you are.

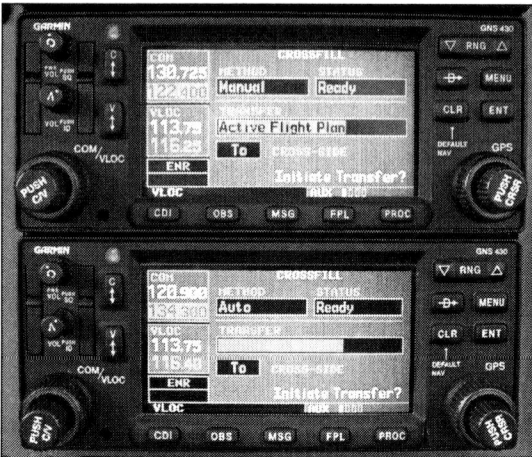

Dual Installations

In an aircraft with two GNS 430 systems (or a GNS 430 and GNS 530) it is very important to know which (if any) is the master and which is the slave - otherwise you can find yourself overwriting active flight plans by entering data into one GPS which overwrites an active plan on the other.

In the AUX Group, select the Flight Planning page and select Crossfill. If both units are set to Auto then the changes to the active flight plan on one will be automatically transferred to the other. If one is set to auto then that unit effectively becomes the master with changes to its plan being copied to the other but not the other way around. If neither is set to auto (or the Crossfill option is disabled) then they effectively work independently (though by using the Crossfill option you can initiate manual transfers).

Unscrambling the GPS

If you are flying a hired aircraft it is quite possible a less knowledgeable pilot has fiddled with the GPS and left it "scrambled", with fields and settings in an unexpected state. This can be especially true of the Map page. To quickly reset it to a reasonable state, consider the following actions:

- Select the Map page (to get to the map page from anywhere press and hold [CLR] until the Default NAV page appears and turn the inner [PUSH CRSR] knob one click clockwise).

- Zoom in or out to show a map scale of about 35nm.

- If there is a -1, -2 or -3 by the map scale it has been "de-cluttered" - press [ENT] briefly until the minus figure disappears.

- If the map still "looks wrong" the defaults may have been changed. Press [MENU] and select Setup Map then press [MENU] again and select Restore All Defaults.

On any other pages that seem wrong, press [Menu] and select Restore Defaults.

Hiring a Plane with a GNS 430 | Key Points to Note

Flight Planning

Flight Plans are entered into the GNS 430 in a very similar way to the way Routes are entered into the portable receivers.

Press the [FPL] button to get up the Flight Planning Catalog and press [Menu] and select Create New Flight Plan. You can enter in the route using the right hand inner and outer knobs. If necessary, press the inner knob to activate the cursor.

When you have entered the plan you may wish to store it so that it can be retrieved in case it is accidentally overwritten. To do this press [Menu] and select Copy Flight Plan and choose a flight plan slot to copy into.

To activate a plan in the Catalog, highlight it and press [menu] then select Activate Flight Plan. This copies it into "slot 0" and it becomes active.

The CDI Button

The CDI button toggles the output to an external CDI or HSI between VLOC mode and GPS mode. The mode is shown just above the CDI button.

In VLOC mode the external device displays indications relative to the VOR beacon tuned in the VLOC frequency and the OBS setting of the external instrument - in other words the GNS 430 and CDI act just like a normal NAV radio and CDI.

In GPS mode the external device displays indications relative to the GPS course (i.e. the current leg of the active flight plan or Goto ("Direct to"). The display is independent of the OBS setting of the external instrument.

Airmanship: When using GPS mode, it is good practice to match the OBS setting of the external instrument to the GPS course to avoid risk of confusion.

The GPS will automatically sequence to the next leg when it reaches the waypoint.

The OBS button

The OBS button can switch the GNS 430 into a mode similar to the OBS mode offered by the portable receivers and described in Chapter 19. Depending on the configuration of the GNS 430 and external equipment it will either prompt on screen for the input of an OBS radial (in which case it operates almost identically to the portable receivers), or it will take the setting of the OBS knob of an external CDI and relay into the GPS for use as the course line.

The OBS Button has additional uses in conjunction with instrument approaches.

Message Prompts

When MSG illuminates above the [MSG] button the message can be displayed by pressing the button. Messages include airspace warnings, the next track and also can include scheduled messages such as reminders to change fuel tank etc.

Other Input Devices

External devices such as storm scopes and even TCAS can have their output added to the GNS430. This normally results in an additional NAV display and/or the superimposition of data onto the map page.

Instrument Approaches

Whilst the GNS 430 is not approved for use for instrument approaches in the UK it provides a very valuable source of situational awareness. The GNS 430 has a full representation of the approaches from the initial approach fix, rather than the partial approaches of the portable models described in Chapter 20.

Approaches (and departures) can quickly be selected and activated using the PROC button. By default the approach to the destination airfield is offered, though you can select other airports.

Approaches are described in immense detail in the GNS 430 manual and it is recommended that you refer to this for more information.

Approaches: Readers are likely to be flying light aircraft (CAT A) and it should be noted that procedure approaches for CAT A aircraft are often "inside" the procedures for heavier aircraft - such differences will normally be noted on the published instrument plate for the airport in question, however such differences are not contained within the GNS 430.

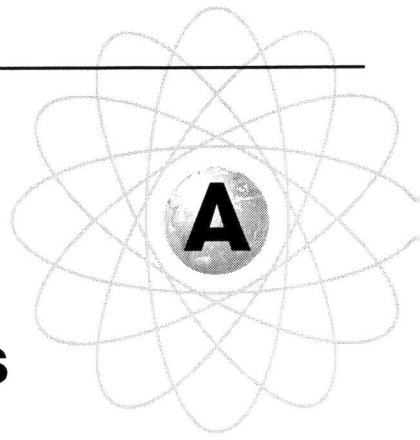

Frequently Asked Questions

General

Q. Is GPS really as vulnerable to failure as scaremongers say?

A. Aside from the factors discussed in Section 1, GPS is very susceptible to interference (some people have compared the signal strength as a light bulb seen at 12,000 miles). Spurious interference from other aircraft equipment is easily possible. The Author has "lost" the GPS for real about half a dozen times, twice due to stretching batteries "one more leg", twice due to route corruption and power offs (early GPSMAP 295), and twice for no apparent reason, once over featureless terrain in the middle of Italy. None of this was a big deal as at all times GPS was a supplement to primary navigation methods.

Q. What is DGPS?

A. DGPS stands for Differential Global Positioning System. It is a method of correcting the inaccuracies of the mobile GPS by using transmissions from a land-based station at a precisely known location. This technique is increasingly being used in surveying, though is not very useful in aviation. This can improve accuracy to 3-5 metres.

Q. What is WAAS?

A. WAAS stands for Wide Area Augmentation System. This can, in theory, improve GPS accuracy to less than 3 metres. It works by ground reference stations at precisely known locations using the existing GPS network to compute their location. They therefore know the inaccuracy and can broadcast a correction signal from correction satellites. At the time of writing there are twenty-five reference stations (all in the US) and there are correction satellites in geo-stationary orbit to give coverage of most of the Earth. The geo-stationary satellites that are most readily receivable in Europe are called AOR-E and AOR-W (For Atlantic Oceanic Region East and West respectively). If you have a GPSMAP 295 with a software version 2.20 or later and you enable WAAS on the Main Menu System tab you should see satellite 33 and/or 35 appear on the Satellite Status page. They correspond to AOR-E and AOR-W respectively. This enables differential WAAS navigation to take place, and you may in the future notice a marked improvement in accuracy of your receiver. WAAS has been partially available since 1999 and is expected to be adopted for active use by the FAA in 2003.

Q. When will we be able to use GPS as a primary navigation aid?

A. This depends in part on the technology but more significantly on the regulatory bodies. The FAA is introducing the use of GPS as a primary navigation aid through the implementation of WAAS to enable en-route navigation, and non-precision approaches in the US. Certified equipment must adhere to the FAA TSO-C129 standard. Many of Garmin's panel mount units are

TSO-C129 compliant, though it is unlikely portable units will meet the requirements. The CAA traditionally takes longer to adopt new technologies!

Configuration Issues

Q. I can't see any NDBs on the map.
A. The default Garmin settings for NDBs mean that you have to be zoomed in a long way for them to appear. For flying in the UK and Europe it would seem better to change the settings as described in Appendix B in order to make them appear more readily.

Q. I can't see small airfields on the map
A. The default Garmin settings for small airfields mean that you have to be zoomed in a long way for them to appear. For flying in the UK and Europe it would seem better to change the settings as described in Appendix B in order to make them appear more readily.

Q. The GPS displays the wrong time
A. The GPS clock is reset very accurately whenever it receives satellites (when it isn't receiving satellites it can gain or lose a small amount of time just like any clock). If the GPS is receiving satellites and still seems to be wrong, the time zone may be wrong. Check and set the time zone as described in Chapter 8.

Q. What is the HSI Bug and what can it do?
A. The HSI Bug is the little triangle around the edge of the HSI display. By default it recommends a course to steer to regain track. You can also set it to indicate the Bearing to the next waypoint or a manual setting using the Set Bug Indicator option on the HSI screen context sensitive menu.

Operational Problems

Q. When I pass a waypoint the GPS doesn't automatically sequence to the next one on my route and tries to guide me back to the last one.
A. This is likely due to one of three reasons. Check that you have a route activated and not just a Goto. Check that you haven't used the Set OBS and Hold function (If necessary Release Hold from the HSI page context sensitive menu). Finally, you may have cut the corner and not got near enough the last waypoint. If that is the case Re-evaluate (Reactivate) the route using the context sensitive menu on the Active Route page - this should make the GPS realise which waypoint you are now heading for (if it doesn't pause a moment and try it again).

Q. The GPSMAP 295 switches off for no apparent reason.

A. Sometimes the GPSMAP 295 will switch off for no apparent reason. The Author has seen this twice for real in the air and quite a few times on the ground (including while preparing this book). Garmin have made a number of changes and this problem appears to be fixed from software version 2.28 onwards.

Q. The GPSMAP 295 seems to have strange characters in its route.

A. The Author has seen this a number of times and it sometimes seems to be associated with the switching off problem mentioned above. The only cure seems to be to delete all the routes. Garmin have made a number of changes and this problem appears to be fixed from software version 2.28 onwards.

Q. Why does the distance shown on the GPS differ from the aircraft's DME?

A. The GPS shows horizontal distance while the DME shows slant distance. In addition if the DME is not co-located with a VOR (e.g. if it is a TDME on an airfield) it may be at a slightly different position than the waypoint in question.

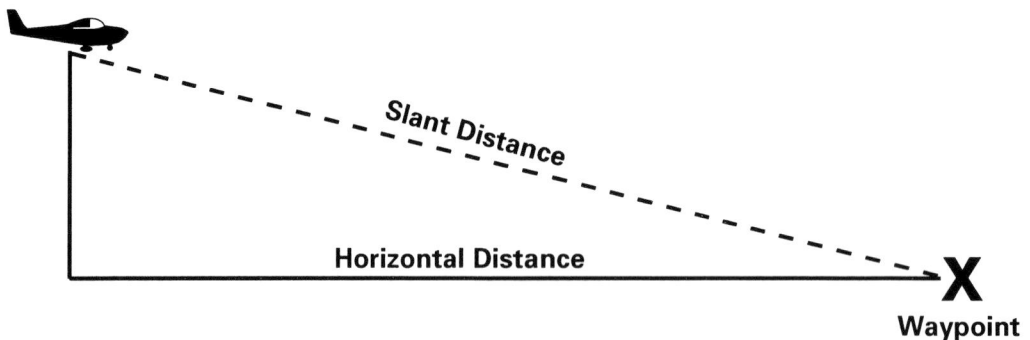

Slant Distance

Horizontal Distance

X

Waypoint

Q. When I pan (scroll) the GPS map manually it keeps pausing.

A. This is normal, as the GPS has to prepare the map data to display. If you want to scroll a long way, it may be quicker to zoom out, scroll and zoom back in. Scrolling is significantly improved on the GPSMAP 196.

Q. I've panned (scrolled) the GPS map and now can't get back to normal operation

A. Just press the QUIT button - this will re-centre the map on the present GPS position.

Q. The HSI seems "blank" with no arrows or bug.
A. The Course setting, D-bar and Bug only appear when there is an Active Route or Goto.

Q. When I have the Map page configured with an HSI, how do I set the Map page HSI sensitivity?
A. The HSI on the Map page will have the same sensitivity as the one on the main HSI page. Set the sensitivity on the HSI page using the zoom in and out buttons.

Simulator Problems

Q. The plane doesn't move on the Simulator.
A. You need to set a speed. You do this using the rocker pad/up down on the HSI page (or editing the speed on the GPS 92 CDI page).

Q. The plane isn't following the planned route on the Simulator.
A. You may have inadvertently started manually steering the plane by using the rocker pad left/right (perhaps when setting the speed). You will need to reinitialise the simulation as described in Chapter 9.

Q. When I try to steer the simulated plane, the map just pans (scrolls).
A. This is normal. The plane can only be steered on the HSI page.

Q. When I've simulated a route the plane just keeps on going in a straight line after reaching the destination.
A. This is normal - unlike a real pilot the Simulator can't do circuits and landings! You can stop the simulator by switching it off as described in Chapter 9 or set the speed to zero to pause it while you plan a new Goto or route.

Q. I switched the GPS off during a simulation exercise - how do I carry on next time?
A. The GPS switches off the Simulator when you turn the GPS off as a safety feature so you don't inadvertently forget and fly with it in that mode. To continue an exercise from where you left off you will have to switch the Simulator back on and re-initialise your position. Approximating your position is normally OK. You may also need to reactivate the route or Goto.

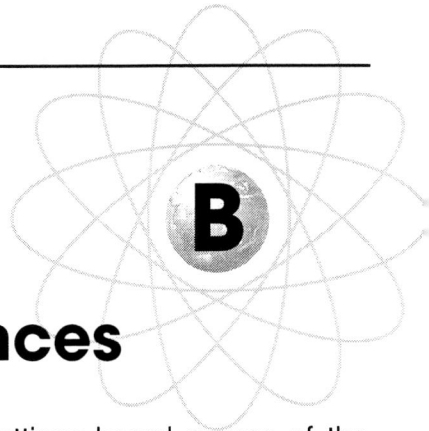

Map Configuration Preferences

This appendix contains the Author's preferences for map page settings based on use of the Garmin portable GPS in the UK and Europe.

The settings tend to give a good general view when zoomed out balanced with the right level of displayable data at the practical zoom levels used in-flight. Any excess map clutter tends to be present only at lesser-used zoom settings.

Map Page Zoom Suggestions

Description	GPS 92	GPS III Pilot	GPSMAP 195	GPSMAP 295
En Route	30nm	8nm	30nm	8nm
Approach WPT	20nm	5nm	20nm	5nm
Approach APT	12nm	3nm	12nm	3nm
Circuit	8nm	2nm	8nm	2nm

Map Feature Setup Suggestions

Description	GPS 92	GPS III Pilot	GPSMAP 195	GPSMAP 295
Map Detail	N/a	N/a	N/a	Normal
Orientation	North Up	North Up	North Up	North Up
Color Mode	N/a	N/a	N/a	Auto
Auto Zoom	No	Off	Off	Off
Land Data	N/a	Gray	Gray	On
Aviation Data	N/a	On	On	On
Track Log	Yes	12nm	80nm	30nm
Active Rte. Line	Yes	200nm	1200nm	800nm
Lat/Lon Grid	N/a	Off	Off	Off
Rings	No	N/a	N/a	N/a
User Waypoint	50nm	Small/8nm	Small/50nm	Small/12m
Symbol Wpt.	N/a	N/a	50nm	N/a
Active Wpt.	N/a	Med/30nm	Medium/200nm	Med/80nm
Large Airport	80nm (APT)	Med/20nm	Medium/120nm	Med/30nm
Medium Airport	80nm (APT)	Small/12nm	Small/80nm	Small/20nm
Small Airport	80nm (APT)	Small/8nm	Small/50nm	Small/12nm
R/W Numbers	N/a	N/a	N/a	Small/3nm
VOR	80nm	Small/12nm	Small/80nm	Small/30nm
NDB	50nm	Small/8nm	Small/50nm	Small/20nm

Map Configuration Preferences   **Setup Suggestions**

Map Feature Setup Suggestions				
Description	**GPS 92**	**GPS III Pilot**	**GPSMAP 195**	**GPSMAP 295**
Intersection	Off	None/Off	None/Off	None/Off
Class B, CTA	120nm	20nm	120nm	150nm
Class C, CTR	80nm	12nm	80nm	30nm
Tower Zone	N/a	12nm	80nm	30nm
Restricted	120nm	20nm	120nm	50nm
MOA	120nm	20nm	120nm	50nm
Mode C Veil	N/a	N/a	N/a	50nm
Other	80nm	20nm	N/a	50nm
Large City	N/a	Med/30nm	Med/200nm	Med/80nm
Medium City	N/a	Small/3nm	Small/20nm	Small/8nm
Small City	N/a	Small/2nm	Small/12nm	Small/5nm
Small Town	N/a	N/a	N/a	Small/3nm
Freeway	N/a	30nm	200nm	80nm
Highway	N/a	3nm	20nm	5nm
Local Highway	N/a	2nm	12nm	N/a
Local Road	N/a	2nm	8nm	3nm
Local Rd name	N/a	N/a	N/a	Small/3nm
Railroad	N/a	3nm	20nm	Small/5nm
GEO	N/a	N/a	N/a	Off/Off
Marine Navaid	N/a	N/a	N/a	Off/Off
Exit	N/a	N/a	N/a	Small/5nm
POI	N/a	N/a	N/a	Small/3nm
State, Prov	N/a	Med/120nm	Medium/800nm	N/a
River, Lake	N/a	Small/3nm	Small/20nm	Small/8nm
Park	N/a	N/a	N/a	Small/5nm
Metro Area	N/a	20nm	120nm	50nm
Other	N/a	N/a	N/a	Small/8nm

Map Configuration Preferences   **Setup Suggestions**

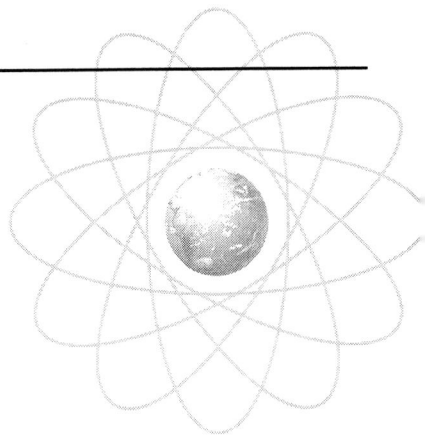